Servois' Solutions

peu connues de différens problèmes de

Géométrie-Pratique

Salvatore J. Petrilli, Jr

Docent Press

Docent Press
Boston, Massachusetts, USA
www.docentpress.com

Docent Press publishes books in the history of mathematics and computing about interesting people and intriguing ideas. The histories are told at many levels of detail and depth that can be explored at leisure by the general reader.

Cover design by Brenda Riddell, Graphic Details.

Produced with TEX.

Illustrations from Servois' *Solutions peu connues de différens problèmes de géométrie-pratique* appear with the kind permission of the Université de Strasbourg, Service Commun de la Documentation (France).

Little-known Solutions

to Various Problems in

Practical Geometry[1]

To serve as a supplement to known Treaties of this Science

Collected by F. J. Servios, *Professor of Mathematics at the Artillery School.*

Nec secius omnis in unguem Arboribus positis secto via limite quadret.[2]

Virgil

---

[1] *Solutions peu connues de différens problèmes de géométrie-pratique*, Paris, Bachelier, 1804.

[2] And no less let every path, or space, square with the cross path, the trees being planted exactly [56].

This work is dedicated to Dr. Robert E. Bradley,
Professor of Mathematics at Adelphi University,
who ignited my love for the history of mathematics.
Rob's superior teaching and scholarly abilities made
this work possible.

Acknowledgments

I am are very grateful to Dr. Robert E. Bradley who offered assistance with translation questions, read and made corrections to the first draft of the translation, and made many helpful suggestions.

Also, I am grateful to Nicole Heyd, Adjointe au responsable du Service de la conservation et de la valorisation, of the Université de Strasbourg, who gave permission for me to reproduce the images from thier copy of Servois' *Solutions peu connues de différens problèmes de géométrie-pratique.* All images have been reproduced with kind permission of the Université de Strasbourg, Service Commun de la Documentation (France).

Finally, I want to thank my father, Salvatore J. Petrilli Sr., who provided me with descriptions of the technical surveying tools utilized in Servois' text.

Salvatore J. Petrilli, Jr.
Garden City, New York
April 2017

Contents

List of Figures

Translator's Preface

a truly original work, notable for presenting the first applications of theory of transversals to the geometry of the ruler or surveyor's staff, thus revealing the fruitfulness and utility of this theory [41, p. 44].

General Overview

In a list of all mathematicians, François-Joseph Servois (1767–1847) would stand to be lesser-known. To the extent that his name is known at all, it is for introducing the words *distributive*, *commutative*, and *pole* to mathematics. However, Servois made significant contributions to mathematics, in particular to the field of linear operator theory.[3]

Servois published his first and only book titled *Solutions peu connues de différens problèmes de géométrie-pratique* (Little-known Solutions to Various Problems in Practical Geometry)[4] in 1804.[5] Servois' geometry text was intended to be a reference on applied geometry for military officers, presenting constructions in ruler-geometry that could be used on the battlefield at any time, which he compiled from his career in the military.[6]

The geometry textbook is divided into two sections after a brief introduction: *Theory* and *Practice*. There are two appendices: a letter regarding questions about second degree equations and an index to the practical section. Then the book ends with an Errata and a plate of figures.

[3]See p. xxiv below for more information about Servois' career in mathematics.

[4]Servois' work is henceforth referred to as simply *géométrie-pratique*.

[5]The year of publication is given as 1805 in [41, p. 44] and [53, p. 325], but is actually year XII of the French Revolutionary calendar, which translates to 1804 or late 1803.

[6]See [10, pp. 2–5] and p. xxi for more information about Servois' military career.

According to Taton, Servois' *géométrie-pratique* was "well-received by the public" [53, p. 325]. However, the question must be asked, who was it "well-received" by? It would appear that Taton is speaking mainly of the mathematical community. There appears to be no evidence stating how this book was specifically used in wartime other than by Servois himself. Servois states:

> If the target does not lie in a straight line, we may use the following method, which I have tried several times successfully [43, p. 63].

What we do know for a fact is that this book brought him fame in the area of projective geometry, which is evident because the great mathematician Jean-Victor Poncelet (1788–1867) consulted Servois for his expertise on geometry several times during the writing of the 1822 edition of the *Traité des propriétés projectives* [53, p. 325].

Finally, much of Servois' work makes use of very basic projective geometry. However, a reader who is not familiar with projective geometry can still make their way through this work. If a reader wishes to gain some background on projective geometry, then *Geometry Revisited* by Coxeter and Greitzer [14] provides a gentle introduction to the topic.

The Contents of the *géométrie-pratique*

Chapter 1: *Introduction*

Servois' *géométrie-pratique* opens with a short four page introduction to his book, which states that this book is intended to serve as an aid to military officers in executing geometric operations on the ground. This was meant to be a very practical and efficient book to be used during wartime, because all these operations could be executed without the use of heavy or expensive surveying tools.

The reader is also provided with the format of his book, which includes a chapter on *Theory* to justify all the operations explained in the chapter on *Practice.* As this stands as Servois' first major work in mathematics, we are exposed to the type of mathematician that Servois would become. Later in his mathematical career, we see that Servois becomes a strong advocate for rigor in mathematics.[7] It is clear that military officers

[7] In mathematics, the term rigor has many definitions. The most appropriate definition to describe Servois' rigor is provided by Grabiner: *First, every concept of the subject had to be explicitly defined in terms of concepts whose nature was held to the already known .... Second, theorems had to be proved, with every step in the proof justified by a previously proved theorem, by a definition, or by an explicitly stated axiom. Third, the definitions chosen, and the theorems proved, had to be sufficiently broad to support the entire structure of valid results belonging to the subject* [21, p. 5]. See p. xxiv for more information about Servois' growth as a mathematician.

would not be interested in the geometric justifications for these operations, but would rather have the simplest steps required to complete these tasks quickly.

Chapter 2: *Theory*

Servois' chapter on *Theory* is intended to give a reader the necessary mathematical background to justify the constructions in his chapter on *Practice.* Servois states that these geometrical theorems are not found in Euclid's *Elements*, but are given by other mathematicians. Servois makes it clear that these theories are not new and that he is not adding original material to the theorems or proofs. However, Servois cites where theorems come from and attempts to organize them in a new way, perhaps, for a non-mathematical reader.

Here are some of the well-known theorems that Servois examines in his chapter on *Theory*:

1. Ceva's Theorem.
2. Menelaus' Theorem.
3. The Construction of the Centroid of a Triangle.
4. The Construction of the Incenter of a Triangle.
5. The Construction of the Orthocenter of a Triangle.
6. The Euler Line Theorem.
7. Desargues' Theorem.
8. The Theorem of Pappus.
9. Monge's Theorem.

Chapter 3: *Practice*

General Practice

Servois' chapter on *Practice* stands as the longest chapter in the book, which is 48 pages long. Servois compiled a list of problems that he felt were most important for military officers to have solutions to, which could be completed with the aid of only Stakes and Chains. All of these constructions rely entirely on the *Theory* chapter. Here are the problems that Servois presents:

1. To prolong a line beyond an obstacle.
2. To find a point aligned with the two invisible points of intersection of two pairs of lines given in direction.
3. From a given point, to draw a line to the invisible point of intersection of two lines given in direction.
4. At a given point, to draw a line parallel to two other parallels given in direction.
5. To draw two parallel lines on the ground.
6. From a given point, to draw a parallel to an accessible line.
7. From a given point, to draw a parallel to an inaccessible line.
8. To divide a line into two equal parts.
9. To divide a line into any number of equal parts.
10. To divide an accessible angle into two equal parts.
11. To divide an inaccessible angle into two equal parts.
12. To erect a perpendicular on a line at a given point.
13. From an inaccessible point, to drop a perpendicular to an accessible line.
14. From a given point, to drop a perpendicular to an inaccessible line.
15. To measure a line, one of whose extremities is inaccessible.
16. To measure an inaccessible line.
17. To measure the distance from a point to an inaccessible line.

All solutions are presented clearly in a step-by-step fashion. At times, some of his constructions become over-whelming, because of the amount of lines of construction required to complete them. However, if a reader has a pencil and paper at hand, they can easily complete the constructions.

French Fortification Terminology

Servois' Tenth Problem[8] utilizes French fortification terminology, which deals with the well-known problem of finding the *capitale d'un bastion* (capital of a fortress). The figure below labels all the points in French fortifications, which is utilized in the military literature. Servois concerns himself with two majors points, the *capital* and the *salient*. The capital of a fortress is the bisector of the salient angle and the salient is an object that is pointing out.

[8] N.° 36 in *Practice*.

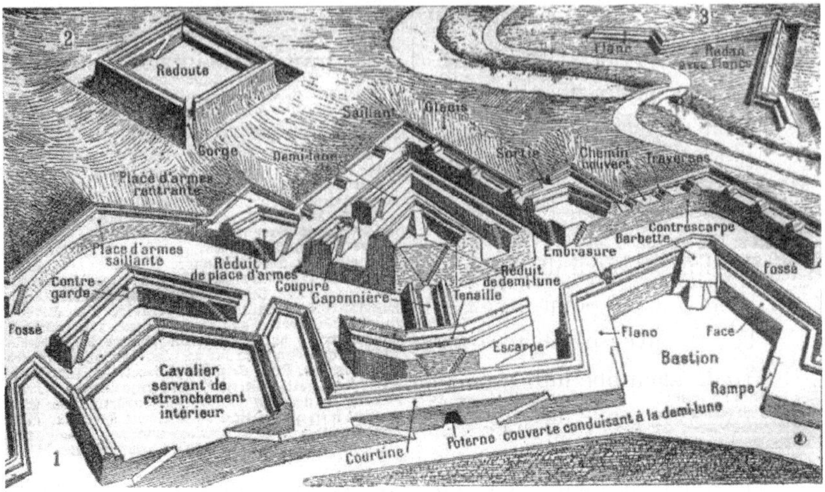

French Fortification Terminology

Little Known Uses of the Équerre d'Arpenteur

Servois' chapter on *Practice* ends with how some of his problems can be solved using two French surveying tools: the Équerre d'Arpenteur and the Fausse Équerre.

The Équerre d'Arpenteur in the following figure is a surveying tool that is used to measure distances from a line. Servois states that there are many variations of this tool and that he prefers the Équerre. This later device can be used to draw vertical or diagonal lines, and verify and draw right angles. However, there are other variations of this tool than can be used to draw angles of any kind. Servois prefers these devices because they can be easily made during wartime. Servois' description of the device offers the ability for a reader to see how such a device can be easily constructed. It should be noted that this device has been retired and replaced by the Double Prism Optical Square.

The Équerre d'Arpenteur is an octagonal cylinder that has slits on each of the eight sides. Half of the openings are rectangular in shape. Each rectangular opening alternates from being at the top of the device to the bottom. Thus, when an observer looks through one of these openings, they are able to define a horizontal plane.

If an observer wanted, for example, to construct a right angle, then they would follow the following steps. You place the Équerre d'Arpenteur at an arbitrary point—call it A. The observer looks through one of the rectangular openings and has a stake placed at some distance in sight—call it B, which can be seen from the device. Then, the observer looks through the next rectangular opening, either to the left or the right depending on the position of the right angle, and has a stake placed at some distance

Équerre d'Arpenteur

in sight—call it C. Therefore, the angle BAC is a right angle. With this basic knowledge, it is clear how to following Servois' directions using the Équerre d'Arpenteur.

The Fausse Équerre in the following figure is another surveying that is used to measure an angle and can be used to transfer an angle. A surveyor can set the arms of the Fausse Équerre to any angle they wish, then place the instrument in a different location and use chains to create a congruent angle in a different location.

Appendix 1: *Letter from S... to F*

Servois' *géométrie-pratique* is also bound with an eight page letter entitled "Letter from S...to F:, Professor of Mathematics," which is most likely addressed to François Joseph Français (1768–1810).[9] This letter has no direct tie to practical geometry. In fact, this letter is about some confusion that Français had on the analytic geometry of second degree equations from a book written by Jean-Baptiste Biot (1774–1862), which is entitled *Traité analytique des courbes et surfaces du second Ordre*. Both questions are about properties of conic sections.

[9]François Joseph Français (1768–1810) and Jacques Frédéric Français (1755–1833) where two brothers that both published in mathematics. Many times in the mathematical literature you will see them simply refereed to as "Français." This letter is most likely a response to F. J. Français, because at this time, J. F. Français took a break from mathematics to pursue his military career.

Fausse Équerre

Due to the content of this letter, it seems odd that Servois would have included this letter as an appendix to his book. However, at this time Servois was working on making a name for himself in mathematics and did not have a venue for publishing his mathematical works. Later Servois would have majority of his mathematical articles and letters published in the *Gergonne Annales des mathématiques pures et appliquées.*

Appendix 2: *Table to the Practical Part*

The final section of Servois' book is a table, which lists all of the constructions presented in the *Practice* chapter. Additionally, Servois provided the section numbers for each construction. It stands to reason that this table was meant to serve as an index for those that needed to find constructions quickly during wartime.

General Remarks

The core of this book consists of the first English translation of Servois' *géométrie-pratique.* There is also a second preface to this work which stands as the most complete biography of Servois available in English. Majority of what we know about the life of Servois comes from [5] and [6].

This works attempts to provide an accurate translation of Servois' text; however, that does not always mean a literal translation. Several of the tools that Servois makes reference to have been left with their French names, because there is no universal name that is used for them. Also, an attempt has been made to simplify Servois' complex sentence structure by breaking longer sentences into smaller pieces.

In Servois' *géométrie-pratique*, the sections were not named chapters, but rather just indicated by their names: *Introduction*, *Theory*, *Practice*, *Letter from S ... to F*, and *Table to the Practical Part.* For clarity, these have been named Chapters and Appendices. Finally, a number within square brackets in the translation indicates the place in *Solutions* where the page with that number began.[10]

Servois' text included a plate with 19 figures. Nicole Heyd, Adjointe au responsable du Service de la conservation et de la valorisation, of the Université de Strasbourg, kindly granted permission to reproduce the images from thier copy of Servois' *Solutions peu connues de différens problèmes de géométrie-pratique.* All images have been reproduced with kind permission of the Université de Strasbourg, Service Commun de la Documentation (France).

Servois' proofs and constructions are relatively straight-forward and clearly presented; however, the most challenging aspect to reading this work would be using Servois' figures to understand the constructions presented. The *Errata* provides some clarification regarding these figures. A reader must envision that Servois' Figures 1a and 1b are a single figure and the same holds true for Servois' Figures 2a and 2b. They have been separated because there would be an excessive amount of construction, which would cause confusion. Servois' Figures 1c, 1d, and 12 are all modifications of Figures 1 and 2. The clarifications noted in Servois' *Errata* make reference to the notion of the *duality* of figures in projective geometry. Duality refers to figures being the same except for position of the points and lines. Finally, Servois has provided two different figures that are labeled as Figure 8 . However, the second one of which he called Figure 8 and shown below is not utilized in the text.

[10]Serovis' *Solutions* never had a second edition published.

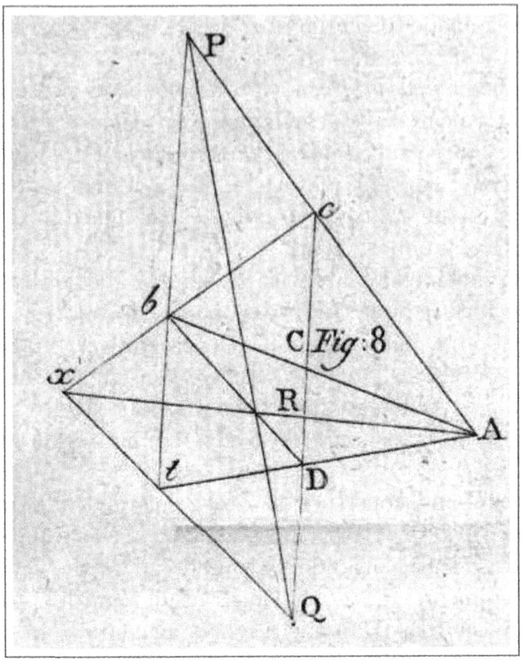

Servois' Unused Figure 8

In this annotated translation, there are several references made to a 2010 work by Aebischer and Languereau titled *Servois ou la géométrie á l'école de l'artillerie* [2]. This is a French text, which provides a biography of Servois and runs through much of the material in the *Theory* and *Practice* chapters from *Solutions peu connues de différens problèmes de géométrie-pratique*; however, they do not discuss the final part of the *Practice* chapter that describe the uses of the Équerre d'Arpenteur. This work contains two very delightful components, which complement this annotated translation. First, the authors provide several classroom activities for several of Servois' constructions, which have been included as footnotes in this annotated translation. In addition, Aebischer and Languereau have created their own diagrams to help understand some of the more difficult constructions. For mathematical clarity, footnotes have been inserted where the Aebisher and Languereau figures aid in understanding Servois' constructions. A reader can use the Aebisher and Languereau figures in one of two ways: use these figures exclusively or use them as a guide to understand how Servois' was handling the issue of duality of his figures. If no footnote appears, then a reader can easily use Servois' figure to understand the construction. This annotated translation does not contain newly created figures but rather present the figures as Servois presented them, which maintains the historical nature of this work.

Besides the classroom activities provided by Aebischer and Languereau many of remaining constructions in Servois' text can serve as enrichment exercises to build spatial and projective skills in geometry. In a study done by Tentomas on spatial reasoning in the classroom [54], it is stated that:

> Evidence from the study suggests that well-designed components can be helpful in improving spatial skills. Evidence from the qualitative results suggested that more time should be devoted to solving spatial ability problems. Unit plans that focus on spatial ability should provide students with problems that require them to use various spatial ability concepts .[11]

Spatial reasoning and projective geometry are intertwined together throughout K–12 geometry curricula. A reader who is interested in more about enrichment projects for geometry of this nature can refer to Tentomas. Finally, this text can serve as a resource for interdisciplinary activities in mathematics, history, and French classes, which is a key component in the Common Core Curriculum for Mathematics.

Salvatore J. Petrilli
Adelphi University
Garden City, NY USA

[11]This quote comes from the unpaginated abstract of Tentomas' dissertation [54].

Translator's Biography of François-Joseph Servois[12]

Family Background and Education

According to his birth certificate (provided by [6, p. 314]), Servois was born on July 19, 1767, in the small village of Mont-de-Laval, located in the Department of Doubs in the Franche-Compté region of eastern France close to the Swiss border. The only son of Jacques-Ignace Servois, a merchant, and Jeanne-Marie Jolliot, Servois did have a sister, about whom little is known except that he spent his retirement years with her [6, p. 324]. It would seem appropriate here to include a portrait of Servois; however, there are no known paintings or photographs of him.[13]

As he grew older Servois attended different religious schools in Mont-de-Laval and Besançon, the capital of Doubs, with the intention of becoming a priest [5, p. 307], and was ordained at Besançon near the beginning of the French Revolution [53, p. 325]. Servois' religious career, however, was destined to be a short one.

Careers

Precisely what religious order Servois belonged to is unclear; however, he was sent to be a teacher of philosophy and curate in Saint-Marcellin [5, p. 306]. Bradley [10, p. 3] assumes that Servois was a Carmelite priest, since sources such as Sorrel [51] state that at the time, the only secondary school for boys in Saint-Marcellin was the Collège des Pères Carmes.

In 1793,[14] Servois left the priesthood and joined the army to become an officer in the Foot Artillery; in many historical texts the Foot Artillery

[12]Majority of this material first appeared in [37] and [39].

[13]See [16, p. 3] for a photograph of Servois' signature, which was provided by Aebischer and Hombeline Languereau [2, p. 22].

[14]Starting on September 22, 1793, all official French records were dated using the new Revolutionary Calendar. All dates in this paper have been converted to Gregorian dates using Martin Liechty's [29] conversion system.

is referred to as the Heavy Artillery. On March 5, 1794, Servois officially entered the Châlons artillery school. He was commissioned as Second Lieutenant in the First Foot Artillery Regiment on November 13 of the same year [5, p. 307].

As Second Lieutenant, Servois was actively involved in battle many times during the revolution. He participated, for instance, in the Crossing of the Rhine and the Battle at Nieuwied (1796–1797). Interestingly, the official French records indicate that Servois participated in the Siege of Maastricht; however, the history of the First Foot Artillery Regiment does not mention them having taken part in the battle. Either there is an error in the records or Servois was sent alone and not with the entire Regiment. In addition, he was posted at the artillery's headquarters in the Army of the Rhine [5, pp. 307–8]. During his time in active duty, Servois showed great leadership abilities and it was recognized when he was appointed adjutant to General Depottre, Chief of Staff for the artillery [5, p. 308].

While on active duty, Servois would spend his leisure time studying mathematics [53, p. 325]. Servois showed his mathematical abilities when he made improvements to one of the cannons, significantly increasing its firing range. Servois suffered from poor health during his years serving as a foot officer. These circumstances gave Servois the desire to engage in a non-active military position, one in academia as a professor of mathematics [5, p. 309]. On March 7, 1800, Servois sent to General Andréossy, the Chief of the Division of Artillery and Engineering in the Department of War, a letter expressing his wishes. A translation of Servois' letter to General Andréossy is given below, which was translated by Bradley [10, p. 4]:

> My General,
>
> I am in a weak state of health, which does not permit me to serve actively without continually counting on the indulgence of my superiors. Additionally, I have a decided taste for the exact sciences, which I have cultivated over the past fifteen years, insofar as I have been able to, along with my other studies and the bustle of military life. This has led me to desire a sort of tranquility into which I may deliver myself whole. Such are the motives that have persuaded me to address myself to you in an application for a vacant position of professor of mathematics at the artillery school in Rennes. My enlightened superiors and my former examiner are most willing to provide you with whatever information you might need about me.
>
> Salutations and respect,
>
> F.-J. Servois.

General Lemaine, brigade commander of the Army of the Rhine, provided a note attesting to Servois' poor health [5, pp. 308–9]. General Lemaine believed a transfer to academia would be more beneficial to Servois and to the French Republic. General Lemaine's letter is given below:[15]

> For more than two years I have been employed with this army. During that time, Servois continually served as a state army officer. I know that he was sick several times and does not enjoy good health. As for his morality, he leaves nothing to be desired.
>
> Zurich, March 16, 1800
>
> Lemaire.

On April 19, 1800, Servois sent a letter requesting a teaching position at the artillery school of Rennes. On the same day, General Hanicuque, successor to General Depottre as commander of the artillery in the Army of the Rhine, wrote a letter of recommendation for Servois' request to be appointed as a professor at Rennes [6, p. 316]. General Hanicuque's letter to General Andréossy is given below:

> General,
>
> Knowing the just spirit that guides you and especially the desire you have to propagate the knowledge and value in the army of the artillery, allow me to recommend to you Servois, Lieutenant to the First Regiment of the Foot Artillery. He seeks a position as a professor of mathematics at one of the artillery schools. This officer has long devoted himself to mathematics, which he has made his principal study, thereby preparing himself for a position of this kind. However, in paying heed to nothing but his zeal, at a time when artillery officer trainees were rare, he took an exam, as a result of which he received his commission and then joined the army, where he has remained since his entry into the corps. He has served in a distinguished manner, and I declare that he has served me in the same way for almost two years, during which I have been chief of staff of the artillery corps of the army. I would prefer, both for the corps and for him, that he be willing to follow a military career, but he tells me, with justification, that it would be more difficult for him to obtain such a position in the future. At the moment, this would provide him with the means to perfect his teaching, which the difficult work of the war forces him to neglect, given that he seeks a position that is presently vacant - the one at Ecole de Rennes. He subjects himself to all the examinations that will be prescribed for him. The citizen

[15] Except where noted, all translations from the French are by the author.

> Lacroix, member of the institute and professor at the school of Polytechnique, will be able to give you the information more advantageously on his [Servois'] talents and morality.
>
> Good day and Respect.
>
> Hanicque.

Servois' experience in the military opened the door for him to begin his work in mathematics. Additionally, General Andréossy was a member of an engineering committee at the start of the revolution and was on the lookout for individuals who had mathematical experience with weaponry [20, p. 592]. It is possible that Andréossy had found out about Servois' talents and felt that having him at an academic French institution would be beneficial to the revolution.

Career in Mathematics

After Servois applied for a position at Rennes, Berthier, the Minister of War, replied to Servois, stating that circumstances had led to the school's temporary closing [5, pp. 308–9]. At the beginning of the revolution, Rennes became a headquarters in the fight against royal insurgents and suffered many severe attacks [42, p. 33]. Thus, it is reasonable to assume that the school at Rennes closed due to damage or a possible evacuation; however, Berthier assured Servois that he would be considered when the school re-opened and at other artillery schools for any similar vacant positions [5, pp. 309–10].

The great mathematician Adrien-Marie Legendre (1752–1833) held many different government positions during the French Revolution [52, pp. 141–46], and one of his responsibilities was assigning professorships at the artillery schools of France. In June, 1801, Berthier sent a letter to Legendre with a list of multiple applicants for positions at the artillery schools. In his response, Legendre stated he would do his best to assign the positions; however, one of the candidates seemed very promising, an officer by the name of Servois [5, p. 317]. Legendre's letter to the Minister of War is given below:

Citizen Minister,

I received, with your letter sent on May 29th, various pieces containing the requests made for the vacant professorships at the artillery schools.

I will make the arrangements called for in this letter right away. However, the information that I will have to collect may cause some delays. I have the honor to inform you, Citizen Minister, that I will inform you of my choices for the positions as soon as I obtain sufficient knowledge of the merits of the candidates. The research I have already done, up to now, I can propose to you, as professor of mathematics at Besançon, the Citizen François-Joseph Servois, Lieutenant in the First Regiment of the Artillery. This officer is not in good health and cannot continue in military service. He desires to be employed as a professor in an artillery school. He has the necessary talents to fill this position with distinction. In addition, he has gained mathematical knowledge during the war and has practiced using the weapons of the artillery. I think, Citizen Minister, that your choice could not possibly fall to a subject better suited to the duties that will be entrusted to him.

Good day and respect.

Legendre.

Paris, June 27, 1801.

During his time in the military, Servois had become acquainted with Sylvestre François Lacroix (1765–1843), who was employed as an additional examiner of candidates for professorships in the artillery corps [10, p. 4]. Early on Lacroix realized Servois' mathematical abilities. Thus, it is not surprising that Lacroix also recommended Servois for a professorship at the artillery schools, which was a high compliment, because Lacroix saw the value in good teachers [52, p. 414–46]. Legendre moved Servois to the front of all the other applicants and gave him a position at Besançon, at an artillery school close to his home village. Ironically, the position that Servois was presented with was once held by Lacroix, before the French Revolution. Bradley [10, p. 5] states that it is unclear what prompted Legendre to place so much faith in Servois. It could have been due to a personal assessment by Legendre, the recommendation from Lacroix, or Servois' own mathematical reputation as acknowledged by General Andréossy. On July 7, 1801, Berthier accepted Legendre's recommendation to appoint Servois as professor of mathematics at Besançon. Berthier's letter is given below:

> In the name of the French Republic one and indivisible, and under the authorization of the first counsel of the Minister of War on the advantageous reports that were done for him on the capabilities and talents of François-Joseph Servois, second lieutenant of the Foot Artillery on foot. He is to be named professor of mathematics at the artillery school of Besançon where a position is currently vacant. At this school, he will serve this army under the direction of the general and superior officers of the artillery. As a consequence of this employment, he will enjoy a salary of three thousand francs per year, as of today. In addition, he will be provided with lodging and such other privileges as pertain.
>
> The Minister of War
> Berthier.

During his first month at Besançon, Servois started creating and organizing classes for what he considered "theoretical instruction" [5, p. 309]. No records indicate precisely when Servois officially began his work on the differential calculus; however, the impetus may have been the possibility of his teaching an analysis course.

At this point, an interesting comparison can be drawn between Servois and Joseph-Louis Lagrange (1736–1813). According to Grabiner, Lagrange was asked to teach a beginning analysis course at École Polytechnique in Paris; however, Lagrange believed that analysis did not belong in a mathematics curriculum. [21, p. 25] He felt that the calculus of his day was not based on mathematical concepts because, at that time, calculus was devoted to the study of velocities, rates of change, and their relationship to time [8, pp. 489–92]. Lagrange stated that a term such as velocity has meaning only in the realm of physics, not in mathematics. In 1797, Lagrange began the movement towards rigor in calculus by publishing *Fonctions analytiques*, an article devoted to the necessity for giving calculus a rigorous foundation. Lagrange [27, pp. 1–6] stated that rigor is needed in mathematics to provide generality for the subject, as well as being a pedagogical tool to build rigorous thinking in mathematics education. Now, the creation of "theoretical courses" could have been a pivotal moment when Servois noticed the necessity of rigor in mathematics, especially in analysis. Therefore, the teaching of mathematics plays a role in igniting the necessity for examining the foundations of calculus.

Servois spent only five months at Besançon before being called back into active duty near Milan, along with the entire staff of the institution. Boyer [5, pp. 308–9] states that during this time Servois had been assigned to the Army of Italy, by General Berthier. In 1801 the Army of Italy was finishing a tough campaign to force the Austrian army into a peace treaty that would benefit the French Republic [25, p. 494]. General Berthier, past commander of the Army of Italy, was also on the lookout, as was

General Andréossy, for artillery officers with mathematical and mechanical backgrounds to aid in his military movements [20, p. 592].

By the start of 1802, Servois received a change of appointment by Berthier and was posted to the faculty at Châlons [6, p. 318]. Servois' response to the Minister of War is given below:

> Citizen Minister,
>
> I have the honor of acknowledging receipt of your order on January 11th, (Office of Personnel of the Artillery School) according to which I must go to the Ecole de Châlons to join the committee as first professor of mathematics.
>
> I will hurry to my new destination.
>
> Good day and respect.
>
> Servois
>
> Plaisance, February 1, 1802.

Châlons was a school very familiar to Servois, since he was a student there. He arrived at Châlons on March 16, 1802 [6, pp. 318–9].

His role as professor at Châlons was short lived. On December 17, 1802, he was posted to the artillery school in Metz; however, this time he was assigned to organize and implement basic geometry courses for surveyors creating communes for the republic [5, pp. 308–9]. According to Bradley, Servois' time at Metz was when his mathematical maturity peaked. Servois was given a lengthy position in academia at the most opportune time, because Napoleon became the Emperor of France in 1804 [10, p. 6]. Napoleon was an advocate and supporter of research the mathematical sciences [20, p. 640], which would benefit the French Republic [52, p. 141]. Napoleon stated, "The advancement and perfection of mathematics are intimately connected with the prosperity of the State" [33, p. 217].

During his time in Metz, Servois wrote and published his first and only book, *Solutions peu connues de différens problèmes de géométrie-pratique* (1803–1804) [*Little-known Solutions to Various Problems in Practical Geometry*], which was devoted to applied geometry. Servois' geometry text was intended to be a reference on applied geometry for military officers, presenting constructions in ruler-geometry that could be used on the battlefield at any time [10, pp.7–8]. In [41, p. 44], Jean Victor Poncelet (1788–1867) declared Servois' work to be,

> ... a truly original work, notable for presenting the first applications of theory of transversals to the geometry of the ruler or surveyor's staff, thus revealing the fruitfulness and utility of this theory [53, p. 325].

In addition, Servois presented his first memoir on the differential calculus to the Académie des Sciences in 1805. This work would not be noticed until it was revised and was re-presented in 1810 [5, pp. 309–10].

On February 25, 1808, Servois was reassigned to the artillery school in La Fère [5, pp. 309–10]. During his time at La Fère, Servois presented three more memoirs before the Académie des Sciences [53, p. 325], one being the revision of his 1805 work on the differential calculus which would eventually become his celebrated "Essai sur un nouveau mode d'exposition des principes du calcul différential." The other two articles, never published, dealt with elements of dynamics and the "determination of cometary and planetary orbits" [53, p. 325]. In 1810, Servois published an article on virtual velocities in the *Mémoires of the Turin Academy* [Ibid.]

Servois had become friendly with Joseph Diaz Gergonne (1771–1859), editor of the *Gergonne Annales des mathématiques pures et appliquées*, while the two men served at the artillery school in Châlons [10, pp. 9–11]. Due to this friendship, Servois had the majority of his work published in Gergonne's journal from 1811 to 1817. It would be through Gergonne's *Annales* that Servois would be introduced to the world and have the opportunity to influence other mathematicians. For instance, the mathematician Jacques Frédéric Français (1775–1833) had the following to say regarding Servois' work on a perpetual calendar:[16]

> The examination of the ingenious triple entry table[17] which appeared in this volume, by Mister Servois, has encouraged me to re-examine in Correspondance astronomique et géographique of the Baron Zach (August 1800) and the article of Mister Gauss who provided him with the idea [18, p. 273].

Servois' most celebrated paper, "Essai sur un nouveau mode d'exposition des principes du calcul différential," his official work on the differential calculus, would appear in 1814. Following Lagrange, his work proposed using power series expansion as fundamental to the differential calculus. He defined the differential operator as an infinite series, which he called an *infinitinôme*, in powers of the difference operator:

$$dz = \Delta z - \frac{1}{2}\Delta^2 + \frac{1}{3}\Delta^3 - \frac{1}{4}\Delta^4 + \ldots.$$

Using this operator, he demonstrated how to find differentials of functions and was able to discover the basic laws of the differential calculus, such as the power and product rules. This paper truly is remarkable because, at the heart of Servois' arguments, lies the use of the linear operator properties in proving theorems about the calculus. Instead of using the terms "linear" or "operator", he used the words *commutative* and *distributive* for the first time in their mathematical sense. Servois stated that the

[16]. See [38] for more on Servois' perpetual calendar.

[17]Servois' "Calendrier perpétuel," [45, p. 88].

Commissioners[18] of the Institut National des Science et des Arts had the following comment about his work on the differential calculus,

> In thereby calling back to the differential calculus several methods, some of which don't seem very appropriate to the current state of analysis, (the author) has done something that is very useful for the science [46, p. 140].

Servois published a second paper on the foundation issues of calculus in 1814, "Réflexions sur les divers systèmes d'exposition des principes du calcul differentiel, et, en particulier, sur la doctrine des infiniment petits" [Reflections on the Various Systems of Exposition of the Principles of the Differential Calculus and, in particular, on the Doctrine of the Infinitely Small]. This paper gives Servois' personal views of the history of calculus and insight into his philosophical positions, particularly on placing calculus on an algebraic foundation rather than that of infinitesimals. Servois warned about the dangers of using the idea of the infinite in teaching analysis to beginners,

> In a word, I am convinced that the infinitesimal method does not nor cannot have a theory, that in practice is a dangerous instrument in the hands of beginners, that it necessarily imprints a long-lasting character of awkwardness and pusillanimity upon their work in the course of applications. Finally, anticipating, for my own part, the judgment of posterity, I dare to predict that this matter will one day be accused of having slowed the progress of the mathematical sciences, and with good reason [47, p. 148].

Servois' main argument was that there is no way to rigorously prove theorems using the idea of infinitesimals. The methods that Servois introduced gave students of mathematics a rigorous foundation upon which to learn calculus. Within this passage, Servois indicates the importance that the teaching of mathematics has on the necessity of rigor in mathematics.[19] Gergonne's *Annales* also provided a source where mathematicians could refute the works of others. Taton states that the Gregonne's *Annales* gave Servois the opportunity to influence other mathematicians regarding the benefits of algebra as a rigorous foundation for mathematics. For instance, [10, pp. 11–13] described a heated correspondence between Servois, Jean-Robert Argand (1768–1822) and Jacques Français (1775–1833) regarding the use of geometry to represent complex numbers, i.e., the Argand Plane.[20]

---

[18]The Commissioners of the Institut were the faculty committee members assigned to referee the mathematical work being presented. Servois stated that Legendre and Lacroix were two members of the second committee to approve the final draft of the "Essai" for publication.

[19]See [11] and [12] for more on Servois' contributions to calculus.

[20]William Rowan Hamilton (1805–1865) in his *Lectures on Quaternions* devotes

Servois' most influential work, "Essai," never received any commentary from other mathematicians of the time period. According to Boyer, "Carried along by the authority of Lagrange's great reputation, the method [calculus through series] now enjoyed a short-lived period of comparative success" [8, p. 260]. Because Servois followed in the footsteps of Lagrange, it may have caused no other mathematician to dare challenge it.

At the start of 1814, Servois was called back into active duty, one last time, to defend Paris from the Austrian and Prussian armies [5, p. 311]. Paris fell on March 30, 1814. Servois played a role in many of the major battles during the Revolutionary and Napoleonic periods. A few weeks, later Servois sent a letter stating that he wished to be sent back to his position at La Fère [*Ibid.*]. Servois' letter to the Director of the Artillery is given below:

> Allow me, my dear general, to rejoice with all the artillery officers, to be returned to my post that I occupied with dignity, at the start of the new academic school year. I am certain that you will believe the sincerity of this testimony and that you will not, as a precaution, request new officers to fill the position. It is true, nevertheless, that I need your help in this matter. I ask to be given special permission to be returned to La Fère. I presented myself to Mr. Peyard, who said to me to await your response because he could not do anything to quicken the response from the artillery. Thus, my dear general, it is from you that I await the order to return to my post at La Fère. You would render me a great service to give me a response that would not take too long, because this place is ruining me.
>
> I am full of gratitude for all that you did for me and I would like the continuation of your kind friendship.
>
> F.-J. Servois
> Hotel Notre - Dame, Paris
> April 14th, 1814.

Boyer includes two similar copies of the above letter; however, they are dated five days apart. [6, pp. 317–19] A short note from Servois, included in the letter given above, is provided below:

> You will find a letter written a few days ago, but which I have not sent before seeing the orders of the notice of appointments posting the new officers.

After Napoleon's loss of Paris in March of 1814, changes were made in the ranks of the superior officers of the French government. Servois'

an entire page of footnotes to Servois, where he states that Servois "with remarkable sagacity foresaw ...the theory of quaternions" with regards to this correspondence [22, p. 57].

original letter was sent to the general, Count Dupont and he acknowledges within the letter, with a tone of uncertainty about the addressee, that there are transitions of power within the new French government. It turned out that, Louis XVIII appointed Dupont as the Minister of War.[21] Therefore, after being notified about the new appointments, Servois addressed the re-written letter to the Director of the Artillery.

When Boyer discovered the above letter from Servois, located in the administrative archives of the French Military, he noticed at the bottom of the letter that the receiving general, who is probably Dupont, had written, in his own handwriting:

> Give him the order to be returned to La Fère.
> D.

After being at La Fère for almost a year, Servois returned to Metz and became acquainted with Jean-Victor Poncelet, also a professor of mathematics at Metz. Poncelet is famous for *Traité des propriétés projectives*, his work on projective geometry. According to Taton, Poncelet consulted Servois on geometry several times during the writing of the 1822 edition of his *Traité* [53, p. 325].

Post Career Activities

Servois' final job was as Curator of the Artillery Museum, located in the 7th Arrondissement of Paris. Currently, it is part of the Museum of the Army. Servois was officially assigned the position on May 2, 1817. However, in Servois' resignation letter, he states that he started the position on June 28, 1816. Either this was a transcription error on Boyer's part, or Servois was acting as curator a year before an official appointment [10, pp. 6–10]. The Duke of Feltre's letter to the Minister of the War Department is given below:

> Minister
> Secretary of State to the War Department
> Under the authorization of the king, François-Joseph Servois, professor of mathematics, has been named Curator of the Artillery Museum in Paris. There he will carry out functions under the orders of the general superior officers and special employees to the central committee of the artillery.
>
> He will receive, as a consequence of the appointment, four thousand eight hundred francs. These funds will be provided by the central committee of the artillery.
> Paris, May 2, 1817
> The Marshal, Duke of Feltre.

[21] This information was provided by Dr. Frederick C. Schneid, Department Chair & Professor of History at High Point University.

On August 17, 1822, Servois was made a Knight of Saint-Louis. The Knights of Saint-Louis was founded in 1696 and later served as a model for Napoleon's Légion d'honneur. Officers were awarded this knighthood for distinguishing themselves through their honor, bravery, and loyalty to the military. Servois was one of 661 officers to receive this honor in 1822 [23, pp. 234–61].

On June 1, 1827 Servois sent his resignation letter to the Minster of War and asked to be allowed to retire:[22]

> My Lord,
>
> I am a sexagenarian, I'm infirm and I'm nearly incapable of even slightly serious work; that is to say, I must give up active service. Consequently, I venture to request to be admitted to retirement effective the coming July 1, 1827. My service details and certificates will be supplied to the office of retirement, but, while awaiting that, here is my brief resume:
>
> F.-J. Servois, born July 19, 1767. Student of the [artillery] school of Châlons, March 5, 1794. Lieutenant in the first regiment of the artillery on foot on November 13, 1794. Professor to the artillery schools on July 7, 1801. Curator of the artillery museum, June 28 1816.
>
> I have the honor of being, with the profoundest respect, My Lord, your very humble and obedient servant,
>
> F. - J. Servois.
>
> Paris, June 1, 1827.

Aebischer and Languereau state that there was an issue with Servois' pension and it delayed his retirement [2, pp. 37–38] Servois received a reply to his resignation letter which stated that there was a conflict over whether his pension would be paid from his administrative positions or teaching positions. Additionally, he was told that this matter could take years to resolve. However, at this time, Servois was suffering from poor health, so he was appointed an aid while this matter was being resolved. Finally, in 1828, Servois was given permission to retire; however, he would only be paid half of his pension until the matter was fully resolved. Servois retired to his hometown of Mont-de-Laval and lived twenty more years. He never married, but retired with his sister and his two nieces [5, p. 313]. On September 29, 1836 a decision regarding his pension issue was finally reached, and he was granted his full pension. Additionally, Servois was made an honorary member of the Academy of Sciences, Literature, and Arts of Besançon on August 24, 1836; however, he never attended any meetings [2, p. 38]. Servois died on April 17, 1847 in Mont-de-Laval [53, p. 325], where, according to Boyer:

> ...*his legacy remains legendary* [5, p. 313].

[22]Partial translation by Bradley [10, p. 7], and remaining by the author.

Chapter 1

Introduction

Little-known Solutions to Various Problems in Practical Geometry

[1] It is useful, above all in War, to be able to execute various geometric operations on the ground, such as *measuring distances*, *making prolongations*, *drawing parallels*, etc., without the aid of pricey Instruments, which one acquires and transports with difficulty. It is therefore useful to present a Collection of Methods where everything is solved with simple *Pickets* or *Stakes*, with *Cords* or *Chains*, or, above all, with the *Équerre d'Arpenteur*,[1] which is an instrument that can be had anywhere, because one can make it anywhere.

It is true that there is no lack, so far, of similar methods. However, it is [2] known, in general, that in operations on the ground, *chaining*, if I may be permitted to call it that, requires a lot of time and care, where as one can always make *Alignments*[2] accurately and quickly. It would therefore be interesting to determine if we could not, through repeated alignments, reduce *chaining* to known processes. It is examining the subject from this point of view that I encountered the Solutions that I offer here.

I have not included in my research all of Practical Geometry. I am lim-

[1]This is a device used to measure distances from a line. This device has several names in surveying, such as a "Cross Staff" or "Surveyor's Square." The French term throughout this work. See p. xiv for more information about this device.

[2]Servois uses the noun *alignment* to mean the line segment and its definite extension.

iting myself to Problems that may be called *Linear.*[3] However, one can easily recognize that our Solutions can also be of great help in other fields.

I divide this Small Work into two parts. In the first, which I call *Theory,* I recall a few Geometrical Theorems of which I need to make applications. These Theorems, not found in the ordinary *Elements,*[4] belong to certain Geometers, whom I am pleased to cite. I have been obliged to repeat a lot of their proofs, because in the original Treatises, these were a part of Theories of greater theories [3] that I could neither cut up nor copy in their entirety. Perhaps I have such good fortune, in my choice of these means, to engage future Authors of the *Elements* to insert those propositions that are the most useful; perhaps they will find that I have generalized some in a novel way.

In the second part, which I call *Practice,* I propose sixteen Problems, which I give the Solutions, immediately indicating the Theorems to which they relate. Sometimes I pause for a moment to compare our Solutions with those found elsewhere. In my means of execution I only use the *Chain* and the *Stake*; presenting the details of the Équerre d'Arpenteur in a separate article. I took care to never use the Chain as a Compass;[5] that is to say, I never determined the position of a point on the ground by the intersection of two arcs, or a line and an arc. Thus, for me the Chain here is nothing but a *Ruler,* and we may regard our series of Problems as an *Essay on the Geometry of the Ruler.*

I call the Solutions that I give *little-known,* but [4] I do not dare to call them *new.* Indeed, I have cited with scrupulous accuracy, when I decide to copy an interesting process from a little-known Work. We will see that I often quote from the Compendium that Mascheroni published in Pavia in 1793, Mascheroni,[6] entitled *Problemi per gli Agrimensori con varie soluzioni*[7] (small quarto 8, 76 pages). I have learned with satisfaction that this Small Work, which contains the most complete Collection that I know of the processes and formulas in all parts of Practical Geometry, has just been reprinted in Paris in French.[8]

---

[3] Servois is not using the term "linear" in the way modern mathematicians use it. In this case, linear refers to presenting ruler constructions.

[4] In [43, pp. 2–4], this title was not italicized. This is a reference to Euclid's *Elements* [17].

[5] I.e., the instrument that can be used for drawing circles or arcs.

[6] Lorenzo Mascheroni (1750–1800).

[7] *Problems in Practical Geometry for Surveryors, with Various Solutions.* This is reference [30] in the current bibliography.

[8] These are references [31] and [32] in the current bibliography.

I close with a brief Statement on the principle uses of the *Équerre d'Arpenteur*; an instrument much neglected in France at present, but better appreciated by Foreigners (witnesses of the *Problemi* of Mascheroni and the Practical Geometry of d'Antoni.[9]), whose principle merit is, as we shall see, to provide the means of indefinitely repeating the same angle (and not necessarily a right angle), a merit that it shares with the *fausse Équerre*,[10] or any instrument that consists of two branches forming any angle between them.

[9]Alexandre Victor Papacino D'Antoni (1714–1786). This is reference [15] in the current bibliography.

[10]This is a device with an adjustable arm that is used to measure out an angle. Many surveyors and contractors call this a Bevel Square, because the term "Bevel" is synonymous with the term angle. The French term will be used throughout this work. See p. xiv for more information about this device.

Chapter 2

Theory

Figures 1.a, 2.a

1.° From a point D, taken in the Plane of the triangle ABC, I drop on the sides opposed to each angle the lines that intersect in a, b, and c.[1] I draw bc that cuts BC at d.

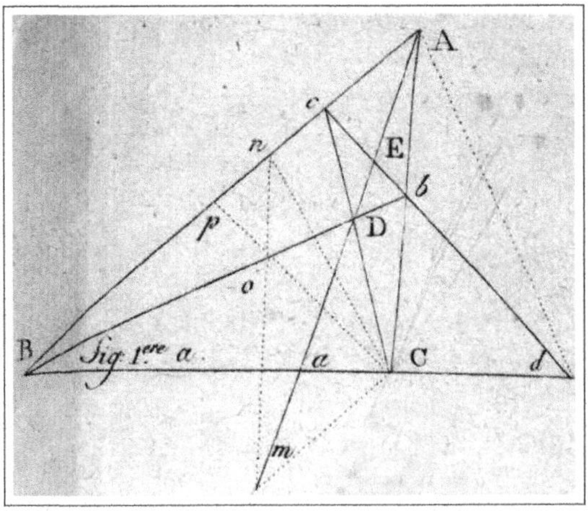

Figure 1.a. Constructing a complete quadrilateral and Ceva's Theorem

Then I draw, through the point C, Cm parallel to AB, and cutting Aa at m. I also draw mn parallel to AC and cutting AB and Bb at n and at o. I join Ad, Cn, and Co. I extend this last one to p on AB. This being so, we first see that Co is parallel to bc, because the two quadrilaterals $CmoD$ and $AbDc$ are equiangular, and they have their

[1]See [2, pp. 64–70] for additional diagrams of Ceva's Theorem as well as ibid., p. 174, for a classroom activity for this passage. A reader can easily see this construction using Figure 1.a alone. However, Servois is also including the case where the point D is outside the triangle in question as in Figure 2.a.

corresponding sides parallel or in the same straight line. Therefore, the corresponding diagonals Co and bc are parallel. It is therefore no less obvious that Cn is parallel to Ad, because they are two corresponding diagonals of the two equiangular quadrilaterals $BnoC$ and $BAbd$, which have their corresponding sides overlapping or parallel.

Through the point d, I draw any transversal $B'd$ cutting AC at C', Aa at a', and AB at B'. I join $B'b$, then I draw $C'q$ at C' parallel to AB until it intersects Aa at q. Then I draw qr parallel to AC until it intersects AB at r, and meeting [6] $B'b$ at s. I join $C'r$ and $C's$, extending the latter until it intersects AB at t. This being so, first of all we have $C'r$ parallel to Cn, because these are two corresponding diagonals of the parallelograms, which have their corresponding sides overlapping or parallel. Therefore $C'r$ is also parallel to Ad. Then we have $C's$ parallel to bc; these are two corresponding diagonals of two equiangular quadrilaterals,[2] having their corresponding sides parallel or in the same line, namely: the quadrilateral formed by the lines $B'b$, $B'd$, qs, and $C'r$, and the quadrilateral formed by the lines $B'b$, $B'd$, AC, and Ad.

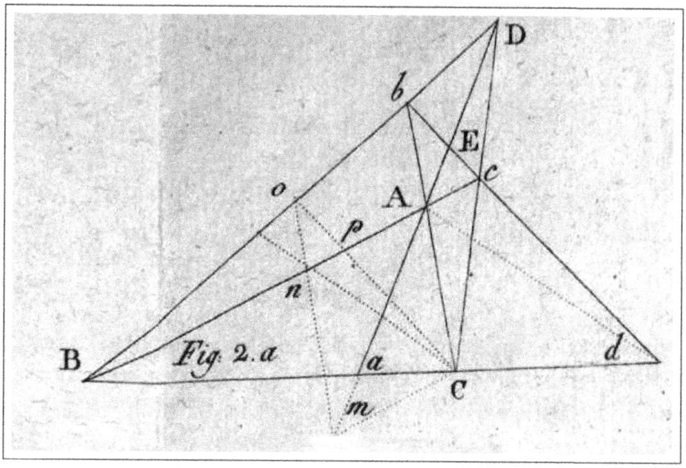

Figure 2.a. Constructing a complete quadrilateral, Ceva's Theorem, and harmonic proportions

2. From this construction,[3] it is without question that

$$Ba : Ca :: AB : Cm \text{ or } An :: Bb : bo :: Bc : cp :: \frac{Bc}{Ac} : \frac{cp}{Ac} :: \frac{Bc}{Ac} : \frac{Cb}{Ab}.$$

Figures 1.ab, 2.a

[2]Quadrilaterals of this nature are called *complete quadrilaterals* in projective geometry. All quadrilaterals in this text are complete.

[3]This is Ceva's Theorem, after Giovanni Ceva (1674–1734). For more information on Ceva's Theorem see [14, pp. 4–5] which also contains a short historical background.

Therefore,

$$\frac{Ba}{Ca} : 1 :: \frac{Bc}{Ac} : \frac{Cb}{Ab} \text{ or } \frac{Ac}{Bc} \cdot \frac{Ba}{Ca} \cdot \frac{Cb}{Ab} = 1. \tag{A}$$

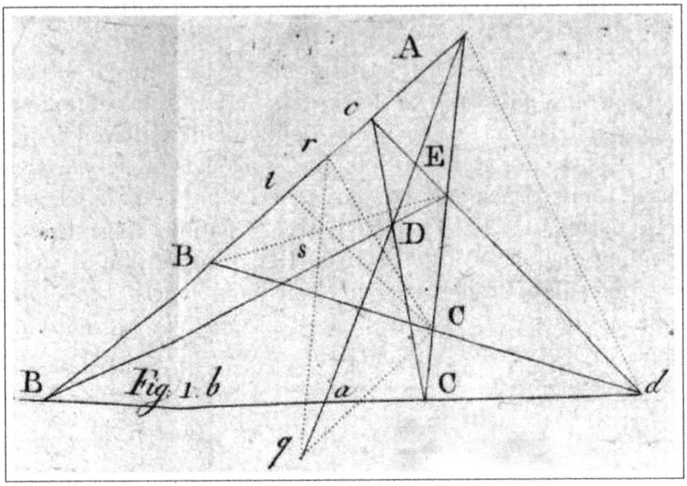

Figure 1.b. Constructing a complete quadrilateral, Ceva's Theorem, and harmonic proportions

Equation (A) written in the form $Ac \cdot Ba \cdot Cb = Ab \cdot Bc \cdot Ca$ expresses this proposition:

> If, from a point taken in the plane of a triangle, we drop straight lines on the sides opposite each angle, we [7] obtain two segments on each side, such that the product of three non-contiguous segments is equal to the product of the three others.

This is the theorem of Jean Bernoulli (Œuvres, volume IV, n. CLV).[4] It is also theorem VII in section IV of *Géométrie de position*[5] of the celebrated Carnot.[6] (Page 281.)

We also have

$$B'a' : C'a' :: AB' : Cq$$

or

$$Ar :: B'b : bs :: B'c : ct :: \frac{B'c}{Ac} : \frac{ct}{Ac} :: \frac{B'c}{Ac} : \frac{C'b}{Ab}.$$

[4]This is reference [3] in the current bibliography. Jean Bernoulli (1667–1748) is also known as Johann Bernoulli. In [43], the spelling was given as 'Bernouilli'.

[5]This is reference [13] in the current bibliography.

[6]Lazare Carnot (1753–1823).

Therefore,

$$\frac{B'a'}{C'a'} : 1 :: \frac{B'c}{Ac} : \frac{C'b}{Ab} \quad \text{or} \quad \frac{Ac}{B'c} \cdot \frac{B'a'}{C'a'} \cdot \frac{C'b}{Ab} = 1.$$

Equation (A) is only a particular case of this, because the line BC is only one of a particular positions of the transversal $B'd$.

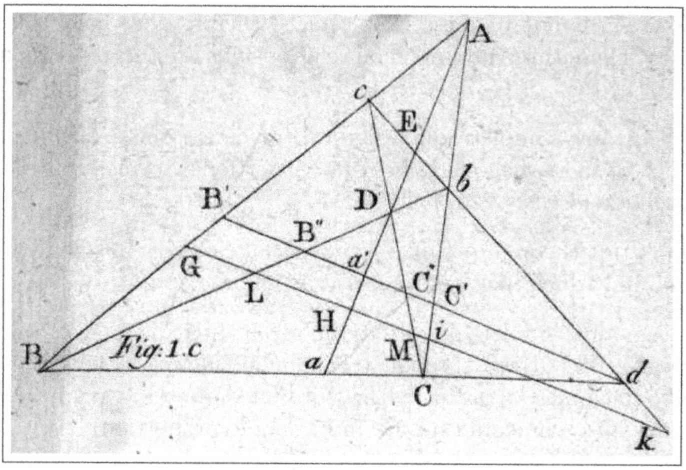

Figure 1.c. Constructing a complete quadrilateral, Ceva's Theorem, and harmonic proportions

3. From the same construction[7] we also have that Figures 1.bc, 2.a

$$B'a' : C'a :: AB' : C'q \text{ or } Ar :: Bd : C'd.$$

Therefore,

$$B'a' \cdot C'd = B'd \cdot C'a'. \tag{B}$$

The preceding proportional relationship can be put in the form

$$B'd : C'd :: B'd - a'd : a'd - C'd.$$

That is to say that the transversal $B'd$ is divided by the lines AC, Aa, and AB into three segments starting from the point d, namely: $B'd$, $a'd$, and $C'd$, which are in harmonic proportion.[8]

[7]See [2, pp. 76-80] for additional diagrams of this theorem. In [43], this proportional relation was given as $B'a' : C'a :: Ab' : Cq$ or $Ar :: Bd : C'd$ but corrected in the *Errata*.

[8]Three quantities are said to be in harmonic proportion, when the first is to the third as the difference between the first and second is to the difference between the second and third.

[8] The lines BC and bc are only particular positions of the transversal $B'd$, therefore we have the proportions

$$Bd : Cd :: Ba : Ca : bd \quad \text{and}^{9} \quad bd : cd :: bE : cE.$$

If we consider the triangle Bcd, and if we imagine a line that joins the points D and d, any transversal through the point A will be cut by the lines bc, Dd, and Bd into segments whose relation is expressed by equation (B). Now, Aa is one of these transversals, therefore we have $Aa : AE :: Da : DE$. These three proportional relationships are the expression of the following theorem of *Géométrie de position.*

> In any complete quadrilateral, given its three diagonals, each of these diagonals is cut in proportional segments by the two others. (N.◦ 225, page 282).

Therefore, this theorem provides only special cases of that given by equation (B), and which may be expressed as follows:

> In any complete quadrilateral, given three diagonals,[10] any transversal drawn through the point of intersection of these two diagonals [9] is cut by the third one, and the two adjacent sides into segments that are in harmonic proportion.

The segments start from the point of intersection.

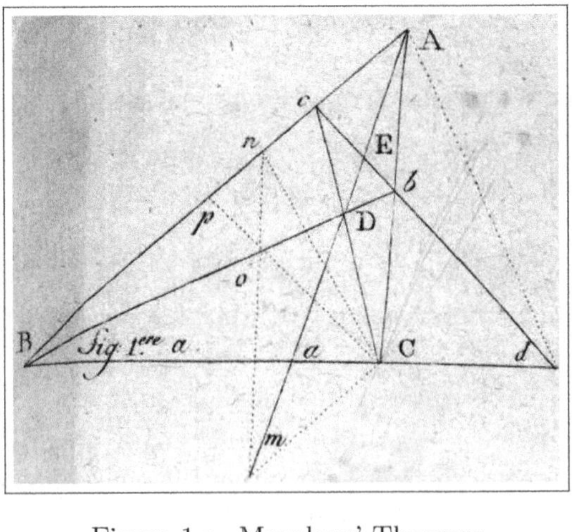

Figure 1.a. Menelaus' Theorem

[9] In [43, p. 8], the following proportional relation was given as $bd : cd :: cE : bE$ but corrected in the *Errata*.

[10] The following footnote was given in [43, p. 8]: "The quadrilateral $AbDc$ in which the points of intersection B and C of the opposite sides are joined by the line BC is, along with the diagonals AD and bc, a complete quadrilateral; BC is the third diagonal."

4. The same construction[11] also gives Figures 1.a, 2.a

$$Ba : Ca :: AB : Cm :: \frac{AB}{Ac} : \frac{Cm}{Ac} :: \frac{AB}{Ac} : \frac{CD}{Dc}.$$

Therefore,

$$\frac{Ba}{Ca} : 1 :: \frac{AB}{Ac} : \frac{CD}{Dc}, \quad \text{or} \quad \frac{Ac}{AB} \cdot \frac{Ba}{Ca} \cdot \frac{CD}{Dc} = 1,$$

or else[12]

$$AC : Ab :: \frac{Cp}{cd} : \frac{bc}{cd} :: \frac{BC}{Bd} : \frac{bc}{cd}.$$

Therefore,

$$\frac{AC}{Ab} \cdot \frac{Bd}{BC} \cdot \frac{bc}{cd} = 1. \tag{C}$$

These two equations do not express two different theorems. By removing fractions, we may translate them by the following general proposition:

> The three sides of any triangle being prolonged indefinitely, if we draw one indefinite transversal that cuts all three, we have on each side two such that the product of three non-contiguous ones among them is equal to product of the other three.

This is theorem V of section IV of the *Géométrie de position.* (Page 276.)

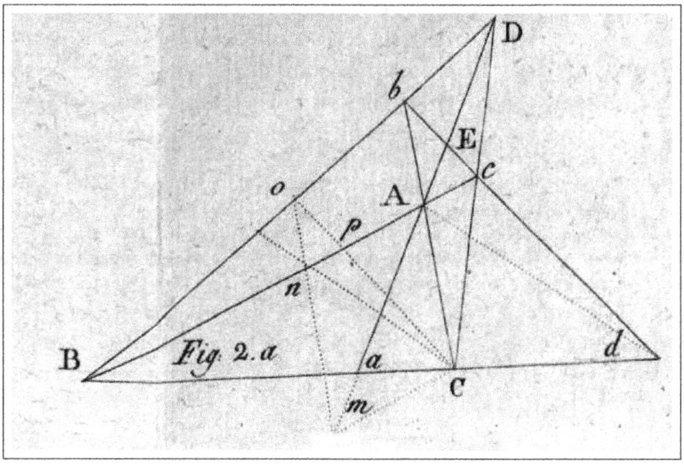

Figure 2.a. Menelaus' Theorem

[11] This is Menelaus' Theorem, after Menelaus of Alexandria (ca. 70–140 CE). See [14, pp. 66–67] for more information on Menelaus' Theorem as well as a short historical background of the theorem.

[12] In [43, p. 9], the proportional relation $AC : Ab$ given as $AC : Ab :: \frac{Cp}{cd} : \frac{bc}{cd} :: BC : bc$ but corrected in the *Errata*. Equation (C) was given as $\frac{Ac}{Ab} \cdot \frac{Bd}{BC} \cdot \frac{bc}{cd} = 1$.

The first equation is given by the triangle BCc and the transversal Aa. The second is given by the triangle Cbd and the transversal AB.

[10] It is not necessary to remark that here by *non-contiguous* segments we mean, as in (N.° 2), those that have no common endpoints. This expression is from Jean Bernoulli:

> *Erunt solida ex tribus laterum segmentis, non contiguis, facta inter se œqualia.*[13]

Figures 1.a, 2.a

5.° The principle lines of figures 1.^{e} and 2.^{e} are equal and similarly arranged; from one to the other I only changed the names of the principle points and the lines of the construction lines that are dotted.[14] Therefore, we may relate the equations found in the preceding articles from the second figure to the first.[15] We will then have the by equation (A)

$$\frac{Db}{Bb}\cdot\frac{Ba}{Ca}\cdot\frac{Cc}{Dc}=1,$$

by equation (B)

$$B''a'\cdot C''d=B''d\cdot C''a',$$

and by equation (C)

$$\frac{CD}{Dc}\cdot\frac{Bd}{BC}\cdot\frac{bc}{db}=1,$$

because from one figure to the other figure, we have $Ac = Db$, $Bc = Bb$, etc. This is only the part of the equations of relation between the lines of figure; my goal is not to present the complete table. The learned author of the *Géométrie de position*, having completed this task in IV section of his book, I turn to consider [11] the many corollaries that can be deduced from the previously demonstrated formulas.

---

[13]The solid figures made from three non-contiguous are equal to one another.

[14]See [2, p. 84] for additional versions of these figures.

[15]In other words, Servois is speaking of the *duality* of his figures 1 and 2, which means that these figures are similar in nature except for the orientation of the points and lines. See p. xviii for more information about the figures presented in [43].

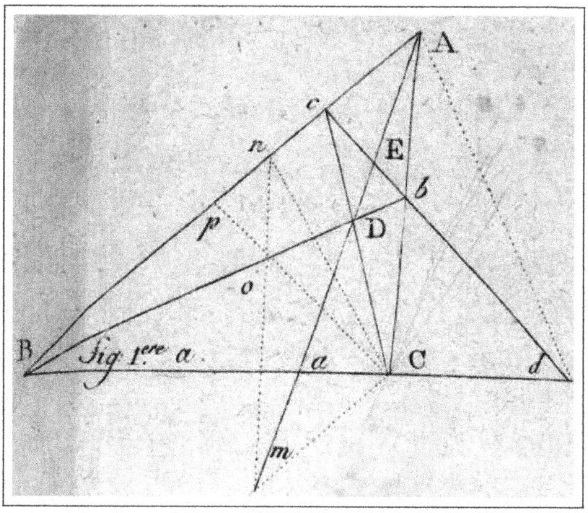

Figure 1.a. The centroid and orthocenter of a triangle

6. Let a, b, and c be the sides opposite of the angles A, B, and C, respectively.[16] If Bb and Cc are perpendicular to the sides b and c, we will have, by a well-known theorem,[17] Figures 1.a, 2.a

$$\frac{Ac}{Bc} = \frac{b^2 + c^2 - a^2}{a^2 + c^2 - b^2} \quad \text{and} \quad \frac{Cb}{Ab} = \frac{a^2 + b^2 - c^2}{b^2 + c^2 - a^2}.$$

Putting these values into equation (2.1), we have

$$\frac{Ba}{Ca} = \frac{a^2 + c^2 - b^2}{a^2 + b^2 - c^2}.$$

That is to say that Ad is also perpendicular on the side a. Thus, the three perpendiculars dropped from the angles of a triangle to the opposite sides intersect at a unique point.[18]

[16]This article makes use of the Law of Cosines, $c^2 = a^2 + b^2 - 2ab\cos(C)$. If a reader keeps this formula in mind, then demonstration made in this article is relatively straight forward. Servois will make use of the Law of Cosines to find the orthocenter of a triangle. This article could make a nice interdisciplinary activity in a mathematics classroom. Also, see [2, p. 86] for an additional diagram for this theorem.

[17]Servois is speaking of the Law of Cosines. In France, this law is called al-Kāshī Theorem after Ghiyāth al-Dīn Jamshīd Mas'ūd al-Kāshī (1380–1429). See [40, pp. 106–7] for more on the history of the Law of Cosines.

[18]This point is called the *orthocenter* of a triangle.

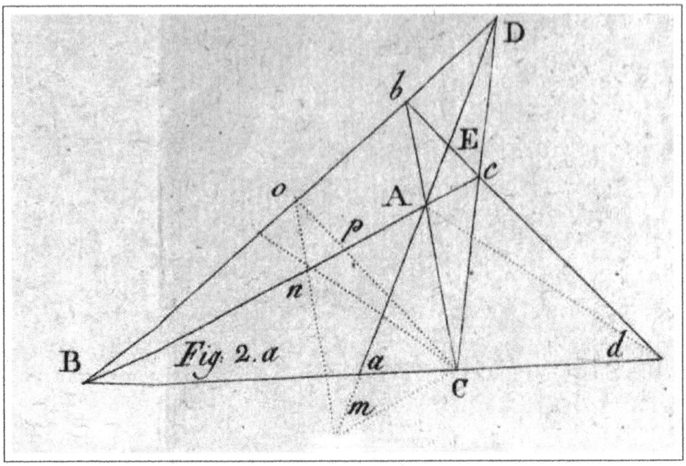

Figure 2.a. The centroid and orthocenter of a triangle

Figures 1.a, 2.a

7. If

$$\frac{Ac}{Bc} = \frac{Ca}{Ba},$$

we have by the same equation (A) $Ab = Cb$, that is to say, if the lines Aa and Cc cut the opposite sides into proportional parts, the line BD always intersects the third side in two equal parts, but then, the line ac is parallel to the side b.[19]

[12] In particular, if $Ac = Bc$ and $Ba = Ca$, this theorem was given as well in *mécanique elementaire.*[20] The three lines dropped from the angles of a triangle to the midpoints of their opposite sides intersect at a unique point. This point is the center of gravity of the triangle.[21]

Figures 3, 4.b

8. If Bb and Cc divide into two equal parts the angles of the vertices from which they are dropped onto the opposite sides, then we will have[22]

$$\frac{Ab}{Cb} = \frac{c}{a} \quad \text{and} \quad \frac{Bc}{Ac} = \frac{a}{b},$$

because if we draw Cm parallel to Bb, then $Ab : Cb :: AB : Bm$. However, the triangle BCm is isosceles, because we have the angles[23] $ABb = BmC$, as corresponding angles; $bBC = BCm$, as alternate interior angles, and $cBb = bBc$ by hypothesis. Therefore, $Bm = BC$ and consequently

$$\frac{Ab}{Cb} = \frac{c}{a}.$$

[19]See [2, p. 89] for an additional diagram of this theorem.
[20]This is reference [19] in the current bibliography.
[21]This point is called the *centroid* of a triangle.
[22]See [2, pp. 92–93] for additional diagrams for this theorem.
[23]In [43, p. 12], the first angle was given as EBb.

Similarly, we may demonstrate that

$$\frac{Bc}{Ac} = \frac{a}{b}.$$

Thus, by equation (A),

$$\frac{Ba}{Ca} = \frac{b}{c}.$$

Therefore, the line[24] Aa also divides the angle from which it originates into two equal parts. Thus, the lines that divide two angles of a triangle into two equal parts, or their supplementary [13] angles, and the line that does the same division on the third angle, intersect at a unique point. This proposition is more general than the well-known theorem stating that the three lines that divide the angles of a triangle into two equal parts intersect at a single point.[25]

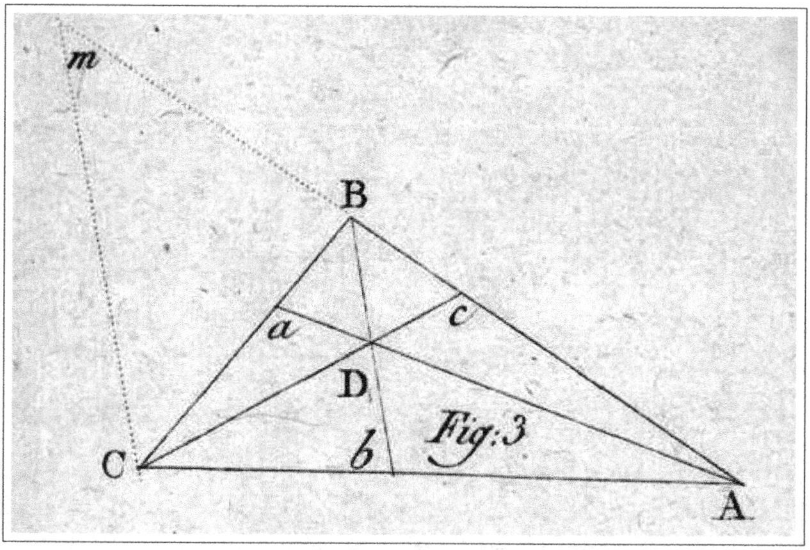

Figure 3. The incenter of a triangle and bisecting a line segment

[24]In [43, p. 12], this line was given as Ad.
[25]This point is called the *incenter* of a triangle.

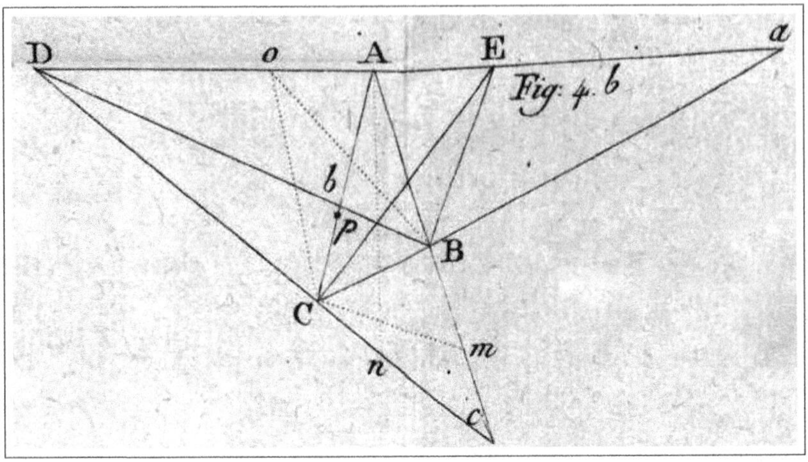

Figure 4.b. Bisecting a line segment

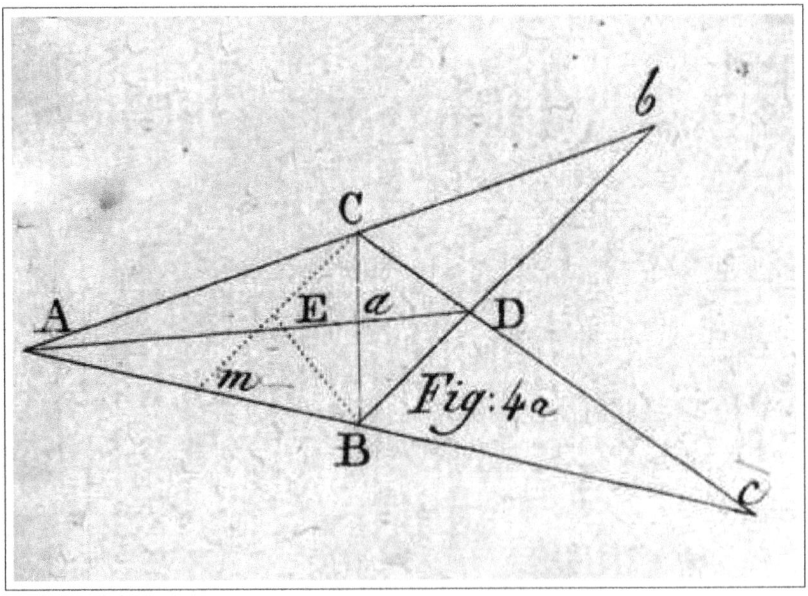

Figure 4.a. Bisecting a line segment

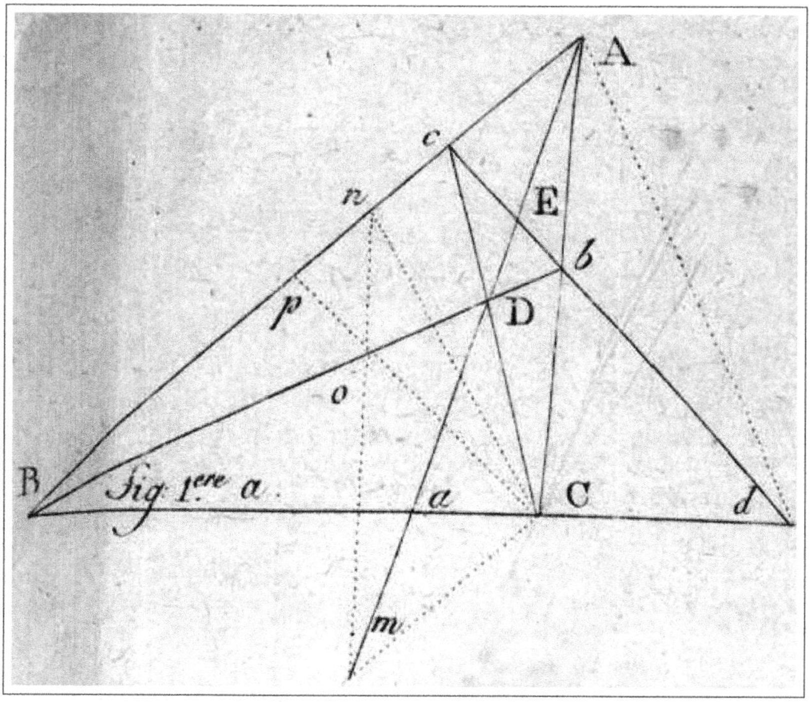

Figure 1.a. Angle and line segment bisections, and a result of Ceva's Theorem

9. If BE and CE are the lines that divide the angles B and C into two equal parts, and BD and CD divide into two equal parts are supplements, then the point E is on AD, and consequently the three points A, E, and D are on the same straight line.[26] Figures 4.a, 4.b

We also see that the lines CD and BD are then perpendicular to CE and BE, respectively.

Under the same hypothesis we have $AD : Da :: AB : Ba$ and $AE : Ea :: AB : Ba$. Therefore, $AD : Da :: AE : Ea$, or $AD \cdot Ea = AE \cdot Da$.[27] This property has been noted by the author of the *Géométrie de position.* N.◦ 197, page 264.)

10. When Figures 1.a, 2.a

$$\frac{Ba}{Ca} = \frac{c}{b},$$

the angle A or its supplement is divided into two equal parts, and we have

[26]See [2, p. 95] for an additional diagram of this theorem. In [43, p. 13], part BD was given as Dd, but corrected in the *Errata*.

[27]$AD \cdot Ea = AE \cdot Da$ was given as $AD - Ea = AE - Da$, but corrected in the *Errata* [*Ibid.*].

by equation (A)

$$\frac{Ab}{Cb} \cdot \frac{Bc}{Ac} = \frac{c}{b}.$$

[14] The latter equation can be verified not only by the hypothesis

$$\frac{Ab}{Cb} = \frac{c}{a} \quad \text{and} \quad \frac{Bc}{Ac} = \frac{a}{b},$$

but by an infinity of others, such as

$$\frac{Ab \cdot Bc}{M} = c \quad \text{and} \quad \frac{Ac \cdot Cb}{M} = b, \quad \text{or} \quad \frac{Ab}{Ac} \cdot M = c \quad \text{and} \quad \frac{Cb}{Bc} \cdot M = b,$$

where M is any line, or etc., etc. That is to say that there is an infinity of ways to divide the sides b and c by lines dropped from the opposite angles, by which the angle A or its supplement will be divided into two equal parts by the line AD.

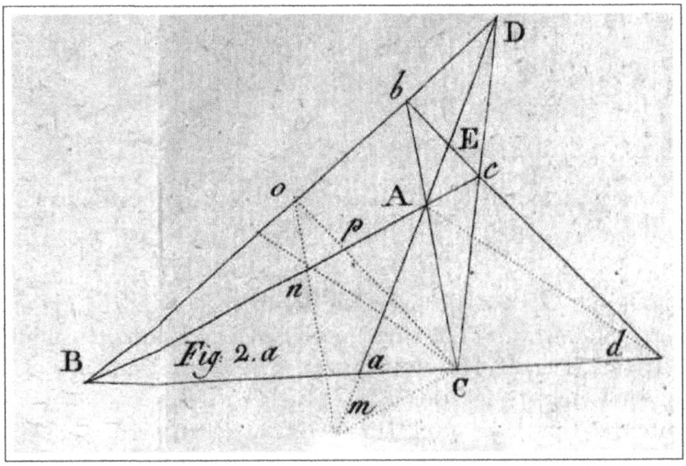

Figure 2.a. Angle and line segment bisections, and a result of Ceva's Theorem

Figures 1.a, 2.a

11. Let

$$Ab = \frac{b}{m}, \quad Ac = \frac{c}{n}, \quad \text{and} \quad Ba = \frac{a}{p},$$

we have by formula (A)

$$p = 1 + \frac{m-1}{n-1}.$$

Thus, from the point A we may draw to BC a line that divides this side into parts that are in a given ratio, even when this side is inaccessible, provided that we can divide the sides AB and AC in such a way to assign m and n values satisfying the indeterminant equation

$$p = 1 + \frac{m-1}{n-1}$$

where p is given.

[15] If $m + n = 2$, we have $p = 0$, and B becomes infinite. That is to say that AD is then parallel to the side BC, but the equation $m + n = 2$ being indeterminant, we see that there is an infinity of possible conditions of division of the sides b and c, under which AD will be parallel to the side a.

12. Let p, q, and r be the midpoints of the three sides of the triangle ABC, D the center of the circumscribed circle, and E the point where the three perpendiculars dropped from the angles to the opposite sides intersect.[28] I join the points p, q, and r with lines, and I have a new triangle clearly similar to the triangle ABC, and for which the point D is the point where perpendiculars dropped from the opposite sides intersect, because the center of the circle circumscribed on the triangle is the point of intersection of the three perpendiculars erected at the midpoints of each side. Thus, D is the point of intersection of the three perpendiculars Dq on AB, Dp on BC, and Dr on AC; however, pr is parallel to AB, pq is parallel to AC, and qr is parallel to BC. Therefore, Dq, Dp, and Dr from the vertices p, q, and r are perpendicular to the sides opposite of them. Now, in two similar triangles, the points of intersection [16] of the perpendiculars dropped from the angles to their opposite sides are similarly placed points. Therefore, the point D in the triangle pqr is placed as is the point E in the triangle ABC. Therefore, $pd : AE :: qr : BC :: 1 : 2$. Joining Ap and DE, we have $pD : AE :: op : Ao :: 1 : 2$. Therefore, op is a third of Ap and the point o is the center of gravity of the triangle ABC, or the point of intersection of the three straight lines dropped from the angles to the midpoints of the opposite sides. Thus, in every triangle, the center of a circumscribed circle, the center of gravity and the intersection point of the three perpendiculars dropped from the angles to the opposite sides are on the same straight line.[29] This theorem was noted by Carnot. (*Géométrie de position*, N.◦131, page 164.) Fig. 5

[28] See [2, pp. 99–100] for additional diagrams for this theorem.

[29] This is the Euler Line Theorem, after Leonhard Euler (1707–1783). See [14, pp. 18–20] for more information on the Euler Line Theorem, which also includes historical background.

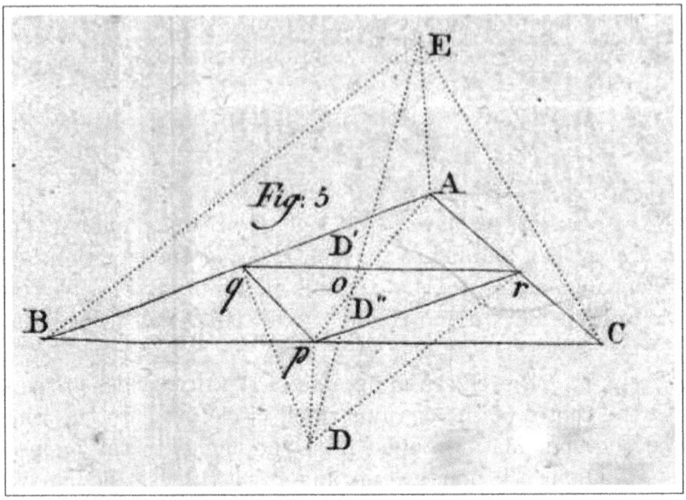

Figure 5. The Euler Line Theorem

The point o is also the center of gravity for the triangle pqr, because it is the intersection of the straight lines qC and pA, dropped from the angles p and q, to the midpoints of the opposite sides. Therefore, the center of the circle that passes through the three points p, q, and r is also to the line DE and is placed with respect to the point D, as the point D is with respect to the point E. Let D' be this point, because we have $D = \frac{1}{2}oE$, we have $oD' = \frac{1}{2}oD = \frac{1}{4}oE$. If we join the [17] midpoints of the sides of the triangle pqr by straight lines to get a new triangle $p'q'r'$, we see that the point o is again its center of gravity, and that the point D'', the center of the circumscribed circle, is again on DE at a distance $OD'' = \frac{1}{2}OD' = \frac{1}{8}OE$ from the point O. If we perform the same operation on the triangle $p'q'r'$ for to get a triangle $p''q''r''$, then on this latter to get the triangle $p'''q'''r'''$, and so on, we see that the centers of all the circumscribed circles of all these triangles are on DE, and that the point o is the limit of all their positions. Therefore, at least for the case of the triangle, we get by purely geometrical considerations the solution of a problem that Ozanam[30] had not believed we could solve without the help of mechanics.

> Having joined the midpoints of the sides of a polygon with straight line, to get a second polygon with the same number of sides, then the midpoints of this one to get a third one, and so on, to find the point where the operation will stop? (*Récréations mathématiques, problème de géométrie*).[31]

[30] Jacques Ozanam (1640–1718).

[31] This is reference [36] in the current bibliography.

13. Let any transversal GK be drawn, cut by the lines AB, AD, Fig. 1.c
AC, and [18] bc at the points G, H, I, and K.[32] From the point d, let the transversal $B'd$ be drawn parallel to GK, we have by formula (B),[33] $B'a' \cdot C'd' = B'd \cdot C'a'$. Furthermore, because of the parallels, we have

$$B'd = Gk \cdot \frac{cd}{ck},\ B'a' = GH \cdot \frac{Aa'}{AH},\ C'a' = HI \cdot \frac{Aa'}{AH}, \text{ and } C'd = IK \cdot \frac{bd}{bk}.$$

Therefore,

$$GK \cdot HI \cdot \frac{cd}{ck} = GH \cdot IK \cdot \frac{bd}{bk}.$$

This equation being applied from the second figure to the first,[34] because of the equivalence of the two figures, it follows that Fig. 1.c

$$KL \cdot HM \cdot \frac{bd}{bk} = HL \cdot KM \cdot \frac{cd}{ck}.$$

Multiply the members of the two equations, we get the unique equation[35]

$$GK \cdot KL \cdot HI \cdot HM = GH \cdot HL \cdot IK \cdot KM. \qquad \text{(D)}$$

This formula is demonstrated in a different way in *Géométrie de position.* (N.◦ 405, page 456.)

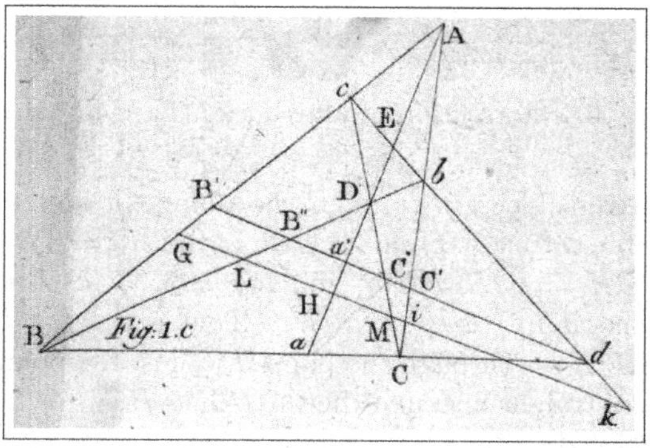

Figure 1.c. Desargues' Theorem

Fig. 6

[32]See figures 2a and 2b in [43] however, those figures are not needed to understand the underlying theory behind this construction. This is Desargues' Theorem, after Girard Desargues (1591–1661). See [14, pp. 70–72] for more information on Desargues' Theorem, which also includes historical background.

[33]In [43, p. 18], this proportional relation was given as $B'a \cdot C'd' = B'd \cdot C'a'$.

[34]See p. 11, footnote 15.

[35]The following equation was given as $GK \cdot KL \cdot HI \cdot HM = GH \cdot HL \cdot IK \cdot KL$. [*Ibid.*]

14. I take AD arbitrarily to the point t. [36] I draw bt and ct until they meet the transveral at L' and M', and I have by equation (D)

$$GK \cdot KL' \cdot HI \cdot HM' = GH \cdot HL' \cdot IK \cdot KM'.$$

[19] Dividing this equation by equation (D) and reducing, we obtain

$$KL' \cdot HM' \cdot HL \cdot KM = HL' \cdot KM' \cdot KL \cdot HM.$$

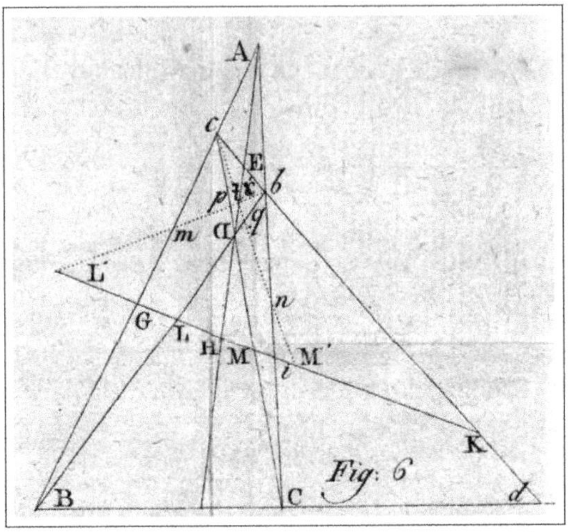

Figure 6. Intersection of four lines at a common point

If the transversal passes through the points m and n on bt and ct respectively intersecting AB and AC, we have $GK = KL' = Km$, $HI = HM' = Hn$, $GH = HL' = Hm$, and $IK = KM' = Kn$, and the first equation becomes $Km \cdot Hn = Hm \cdot Kn$. That is to say (N.◦ 3), that the transversal also passes through the point d. Thus, the points m, n, and d are on the same straight line, and the three lines BC, mn, and bc intersect at the same point d.

If the transversal passes through the points p and q where bt and ct respectively intersect Cc and Bb, we have $KL' = KM = Kp$, $HM' = HL = Hq$, $HL' = HM = Hp$, and $KM' = KL = Kq$, and the second equation becomes $Kp \cdot Hq = Hp \cdot Kq$. Therefore (N.◦ 3), the transversal also passes through the point d, and consequently the points d, p, and q, are on a straight line, and the four lines BC, bc, mn, and pq, intersect at the same point d.

Fig. 7 15. I take the point t at will on AD and the point x on bc. I draw bt, tx, and Ax up till the transversal at L', M', and G', and I have [20] by

[36]See [2, pp. 105–10] for additional diagrams for this theorem.

equation (D), $G'K \cdot KL' \cdot HI \cdot HM' = G'H \cdot HL' \cdot IK \cdot KM'$. Combining this equation with equation (D) to make the factors HI and IK disappear, I obtain[37]

$$G'K \cdot KL' \cdot HM' \cdot GH \cdot HL \cdot KM = G'H \cdot HL' \cdot KM' \cdot GK \cdot KL \cdot HM. \quad \text{(E)}$$

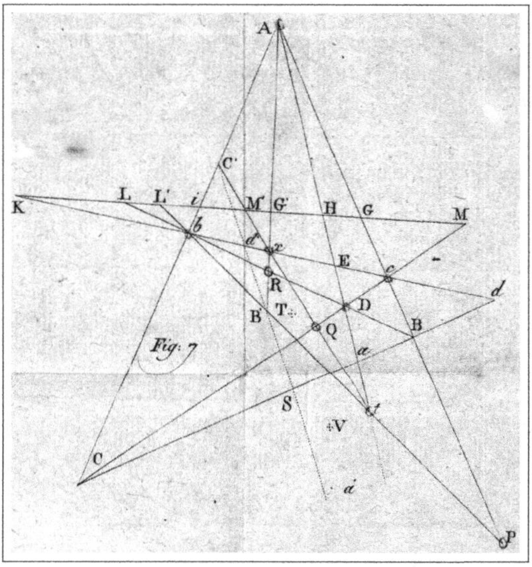

Figure 7. Theorem of Pappus

If the transversal passes through the points P and Q of intersection of AB and bt, and of Cc and tx, we have $HL' = GH$, $HM' = HM$, $KL' = KG$, and $KM' = KM$. The equation (E) becomes $G'K \cdot HL = G'H \cdot KL$ or

$$\frac{HL}{G'H} = \frac{KL}{G'K} = \frac{HK + KL}{HK + G'K},$$

an equation that can not be true except in the case of $KL = G'K$, or in $HK = 0$. In the first case, the point K is the same as the point R where Ax and Bb intersect. In the second case, the transversal passes through the point E where AD and bc intersect, the equation (E) is an identity and shows nothing. However, when tx and Ab intersect at C', and Ax and bt intersect at B', if we imagine the straight line $B'C'$, and if the transversal cuts BC at S, $B'C'$ at S', and Ab at N, we have (N.◦ 3), $EQ \cdot NS = QS \cdot EN$ and $EQ \cdot NS' = QS' \cdot EN$. From this we conclude,

$$\frac{NS}{NS'} = \frac{QS}{QS'} = \frac{NQ + QS}{NQ + QS'},$$

[37] This is Pappus' Theorem, after Pappus of Alexandria (ca. 290 CE–ca. 350 CE). See [14, pp. 67–70] for more information on Pappus' Theorem, which also includes historical background.

an equation that [21] requires that we have $QS = QS'$, because NQ is not zero. Therefore, the points S and S' coincide.[38] This being so, and that the transversal meets Db at R and Ax at R' we also have (N.◦ 3) $EP \cdot RS = PS \cdot ER$ and $EP \cdot R'S = PS \cdot ER'$. Therefore,

$$\frac{RS}{R'S} = \frac{ER}{ER'} = \frac{ES + ER}{ES + ER'},$$

and consequently, ES not being zero, $ER = ER'$. That is to say that the points R and R' concur, and the transversal also passes through the point of intersection R of Bb and Ax. Therefore, the points P, Q, and R are always on the same straight line.

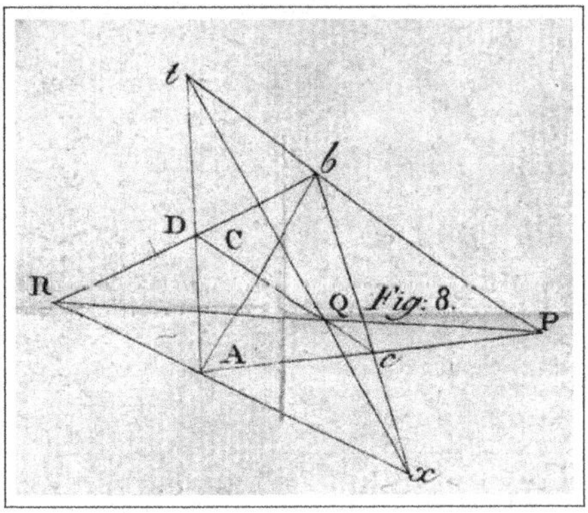

Figure 8.a. Theorem of Pappus

Figures 7, 8 From this, we conclude the following theorem:

> If two quadrilaterals (such as $AbDc$ and $Abtx$) [39] have a common side (such as Ab), and [22] their two diagonals superimposed respectively (such as Ad on Bt and bc on bx), the opposite sides, taken from one quadrilateral to the other (such as bD and Ax, Ac and bt, and Dc and tx) intersect on the same straight line.

[38]The following footnote was given in [43, p. 21]: That is to say, the point S, where BC and $B'C'$ intersect, the point Q where Dc and tx intersects, and the point E where the diagonals bc and Ad intersect, are always on the same straight line; a very remarkable property and worth collecting in passing. We find as well, in drawing Dd, and the line $d't$, form the point d' where $B'C'$ meets bc, the point of intersection T of Dd and td' is, always in straight line with the points A and S. If we draw ac and (through a' where $B'C'$ meets AD) the line $a'x$, the point V of intersection of these two lines, and the points b and S, are always on the same straight line.

[39]Refer to Figure 7 for the quadrilateral $AbDc$ and Figure 8 for the quadrilateral $Abtx$.

Five lines arbitrarily placed in a plane always give two quadrilaterals placed as required by the theorem. These are the five lines: Ab, Ac, Db, Dc, and tx.

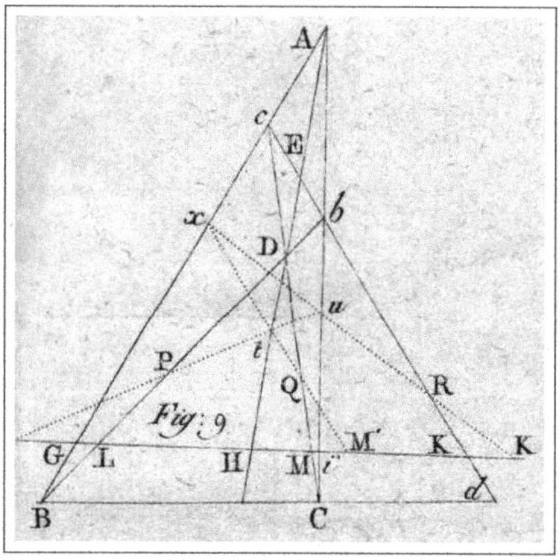

Figure 9. Three points on the same straight line

16. I arbitrarily take the points t on AD, u on Ab, and x on Ac.[40] Fig. 9
I draw tu, tx, and ux up to the transversal at L', M', and K', and I have, by equation (D), $GK' \cdot KL' \cdot HI \cdot HM' = GH \cdot HL' \cdot IK' \cdot KM'$. Combining this equation with equation (D), to remove the factors GH and HI, I obtain:

$$G'K' \cdot K'L' \cdot HM' \cdot HL \cdot IK \cdot KM = HL' \cdot IK' \cdot K'M' \cdot GK \cdot KL \cdot HM. \quad \text{(F)}$$

If the transversal passes through the points P and R, where tu and Bb, ux and bc intersect, we have $GK' = GK$, $K'L' = KL$, $HL = HL'$, $IK = IK'$, and $K'M' = KM'$, and the equation (F) becomes $HM' \cdot KM = HM \cdot KM'$, or rather [23]

$$\frac{HM'}{HM} = \frac{KM'}{KM} = \frac{HK - HM'}{HK - HM'}.$$

An equation that can only hold in the case of $HM' = HM$, or in the case of $HK = 0$. In the first case, the points M and M' coincide, and the transversal passes through the point Q of intersection of tx and Cc. In the second case, it passes through the point E and concurs with the line bc, which cannot happen unless the points u and x are distinguished from the points b and c. Therefore, in all cases, the three points P, Q, and R are on the same straight line.

40See [2, pp. 118–19] for additional diagrams for this theorem.

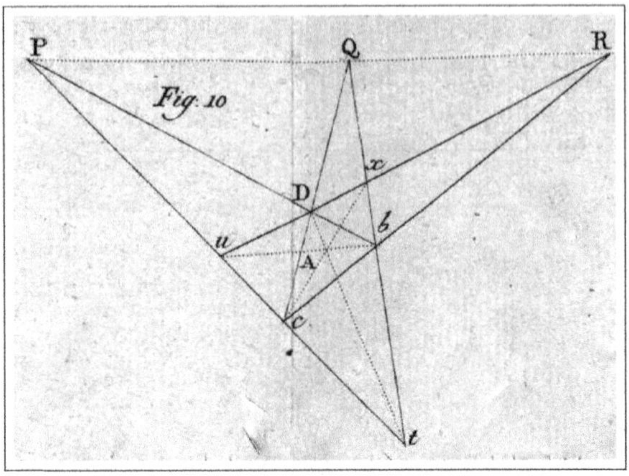

Figure 10. Desargues' Theorem and three points on the same straight line

Figures 9, 10 We see that the two triangles[41] bDc and utx have their vertices b and u, D and t, and e and x, in pairs on three straight lines which intersect at the point A. Thus, we have the following theorem:

> If two triangles are situated so that the three lines that join their vertices taken in pairs from one to the other, such that they intersect in a single point, the three points of intersection of the sides taken in pairs from one to the other triangle opposite the adjacent vertices (that is to say placed on one of the three straight lines that intersect), will always be on the same straight line.[42]

We could also express this theorem in another way:

> If two quadrilaterals (such as $Abdc$ and $Autx$) have one diagonal and the two adjacent sides superimposed (such as Ac, Ad, and Ab superimposed on Ax, At, and Au), the other sides and adjacent diagonals taken from one to the other quadrilateral (such as bc and ux, Dc and tx, and Db and tu), intersect on the same straight line.

Conversely, the points P, Q, and R being on the same straight line, if we arbitrarily draw two lines from each, namely: Pb and Pu from P, Qx and Qc from Q, and Ru and Rb from R, and if we then denote the points b, c, u, and x, where Rb meets Pb and Qc, and where Ru meets Pu and Qx, then the point D, where Pb intersects Qc, and the point t, where Pu

[41] In [43, p. 23], the first triangle was given as bdc.

[42] The following footnote was given in [43, p. 23]: Thus the three points A', B', and C' of figure 86, plate IV of *Géométrie de position*, are in a straight line.

interests Qx, I say that the straight lines bu, cx, and Dt intersect in the same point A.[43]

Indeed, from the point of intersection A of bu and Dt, and through the point c, I draw Ac which cuts the line PQR at the point G. The line PQR is also cut by Dt at H, and by bu at I. This being so, we have (N.◦ 13) $GR \cdot PR \cdot HI \cdot HQ = GH \cdot HP \cdot IR \cdot MR$.

[25] Through the point A and the point x, I draw Ax, which cuts the line PQR at G', and I have (N.◦ 13) $G'R{\cdot}PR{\cdot}HI{\cdot}HQ = G'H{\cdot}HP{\cdot}IR{\cdot}MR$. Comparing these two equations, I obtain

$$\frac{GR}{GH} = \frac{G'R}{G'H} = \frac{GG' + GR}{GG' + GH},$$

an equation that can only hold in the case of $GG' = 0$, because GR is not equal to GH. Therefore, the two lines Ac and Ax coincide, and consequently cx is directed towards the point of intersection A of the two other straight lines bu and Dt.

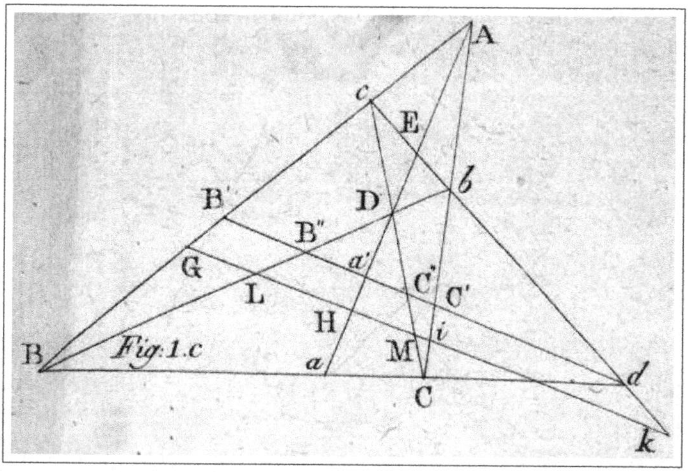

Figure 1.c. Monge's Theorem

17. Equation (C)[44] of (N.◦ 4) coordinates among six points $A, B, C, b, c,$ and d arranged on the four straight lines AB, Bd, AC, and cd, so that the particular combinations of these four points taken three by three, namely:[45] AcB, BCd, ACb, and bcd, are each in a straight lines. From this we may conclude that the six points: $A, B, C, b, c,$ and d being so distributed on the three straight lines Bd, AC, and cd that we have, by Fig. 1.c

[43] See [2, p. 175] for a classroom activity for this passage.

[44] This is Monge's Theorem, after Gaspard Monge (1746–1818). See [2, pp. 120–24] for a modern and detailed explanation of this theorem.

[45] In [43, p. 25], the third straight line was given as *Abc*.

equation (C), that the three points A, c, and B are on the same straight line. Indeed,[46] if the point c, for example, was not on the line AB, the straight line bd would be cut by AB at a point C', drawing Cc' and Bb that intersect at D, [26] we have, by formula (C),

$$\frac{AC}{Ab} \cdot \frac{Bd}{BC} \cdot \frac{bc'}{c'd} = 1.$$

Comparing with the equation

$$\frac{AC}{Ab} \cdot \frac{Bd}{BC} \cdot \frac{bc}{cd} = 1,$$

true by hypothesis, we have[47]

$$\frac{bc'}{c'd} = \frac{bc}{cd} = \frac{cc' + bc}{cc' + cd},$$

an equation that cannot be true, at least if cc' is not equal to zero, because bc is not equal to cd. Therefore, the point c' coincides with the point c, and consequently that point is on the line AB.

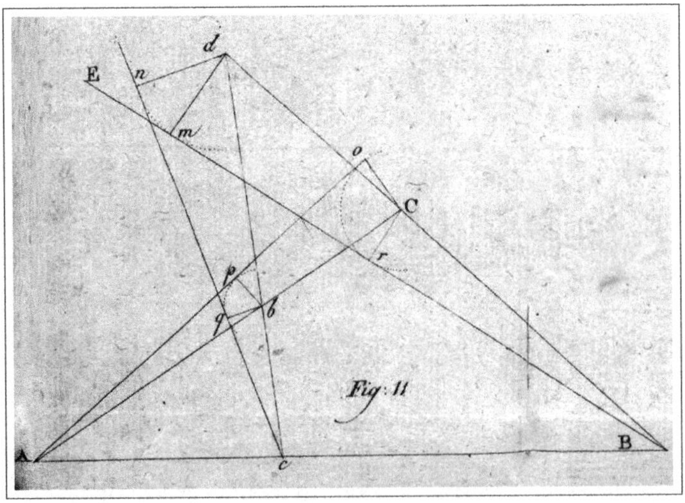

Figure 11. Monge's Theorem

Fig. 11 Let the points b, C, and d be the centers of three circles of any radius. I draw the common tangent mr to the two circles C and d, meeting Cd at

[46] At this point, it seems as if Servois is still making reference to figure 1c; however, he is doing a sort of proof by contradiction by showing that no such construction is possible.

[47] In [43, p. 26], the following formula was given as $\frac{be'}{c'd} = \frac{be}{cd} = \frac{cc'+be}{cc'+cd}$.

B, I draw nq tangent to the two circles b and d, meeting bd at c. Finally I draw op tangent to both the circles b and C, meeting Cb at A. This being so, after having drawn from the points of contact the radii dm, dn, Cr, Co, bp, and bq, we have the three proportional relation $Bd : BC :: dm : Cr$, $AC : Ab :: Co : bp$, and $bc : cd :: bq : dn$, from which we conclude [27]

$$\frac{Bd}{BC} \cdot \frac{AC}{Ab} \cdot \frac{bc}{cd} = \frac{dm}{Cr} \cdot \frac{co}{bp} \cdot \frac{bq}{dn}.$$

However, $dm = dn$, $Cr = Co$, and $bp = bq$, therefore,

$$\frac{Bd}{BC} \cdot \frac{AC}{Ab} \cdot \frac{bc}{cd} = 1,$$

and consequently the points A, c, and B are in a straight line. This is the theorem of Monge[48] (*Géométric descriptive*, N.◦ 44),[49] demonstrated independently of any consideration of projections.

We must remark here that, if the from the point d we erect on Bd two equal straight lines inclined between themselves in any way, namely: dm and dn, and if we draw the straight lines mB and nc, and then from the point C we draw Cr parallel to dn, and Co directed as we may wish, but equal to Cr, and finally from the point b we draw bq parallel to dn, and bp parallel to Co, but equal to bq, then the line op concurs with Cb at the point A on the line Bc. This is because the construction clearly gives the same proportional relationships as above, and therefore the same result.

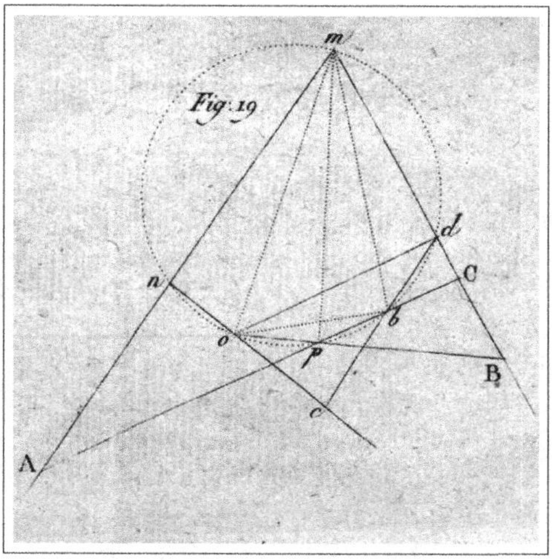

Figure 19. Monge's Theorem

[48] In other words, given three circles in a plane, where none is inside one of the others, the intersection points of each of the three pairs of external tangent lines are collinear.

[49] Monge's *Géométric descriptive* is reference [34] in the current work.

Fig. 19 Let $bdmnop$ be any hexagon inscribed in a circle, and whose opposite sides dm and op, bd and on, bp and mn, prolonged, intersect at the points B, c, and A respectively. Having prolonged Ab until it meets [28] Bd at C, I having drawn the cords bm, bo, and do, we see, without discussion, that we have[50]

$$\frac{AC}{Ab} = \frac{Cm}{bm} \cdot \frac{\sin(dmn)}{\sin(bmn)}, \quad \frac{Bd}{BC} = \frac{do}{Cp} \cdot \frac{\sin(dop)}{\sin(bpo)},$$

and

$$\frac{bc}{cd} = \frac{bo}{do} \cdot \frac{\sin(bon)}{\sin(don)}.$$

Multiplying these equations, and noting that in any quadrilateral inscribed in a circle the opposite angles are supplements of one another, and thus that $\sin(dmn) = \sin(don)$, $\sin(bmn) = \sin(bon)$, and $\sin(bpo) = \sin(bmo)$, we obtain the equation

$$\frac{AC}{Ab} \cdot \frac{Bd}{BC} \cdot \frac{bc}{cd} = \frac{Cm \cdot bo}{bm \cdot Cp} \cdot \frac{\sin(dop)}{\sin(bmo)}.$$

However, in the triangle Cmp, we have

$$\frac{Cm}{Cp} = \frac{\sin(bom)}{\sin(dop)},$$

and in the triangle bom, we have

$$\frac{bo}{bm} = \frac{\sin(bmo)}{\sin(bom)}.$$

Therefore,

$$\frac{Cm \cdot bo}{Cp \cdot bm} = \frac{\sin(bmo)}{\sin(dop)},$$

from which it follows that the right hand side of the above equation[51] reduces to 1, that the same equation becomes equation (C), and that, consequently, the three points A, c, and B are in a straight line.

[29] This is a particular case of the beautiful theorem XLV of the *Géométrie de position.* (Section VI, page 452.)

18. Here I conclude this exposition of lemmas, which I could easily could have extended if I had given in to the appeal offered similar research: from triangles and quadrilaterals we pass without difficultly to polygons of a greater number of sides, and to curves; from plane polygons we pass to polyhedra, where we discover in the intersections of lines and planes, properties for which which I have given above are only particular cases. For example, we find this proposition:

[50]In [43], the notation Sin.A was used to represent the sine of the angle A. The Author is using the modern notation $\sin(A)$ throughout this text.

[51]In other words, the equation $\frac{AC}{Ab} \cdot \frac{Bd}{BC} \cdot \frac{bc}{cd} = \frac{Cm \cdot bo}{bm \cdot Cp} \cdot \frac{\sin(dop)}{\sin(bmo)}$.

> If from the four vertices of any tetrahedron we drop, through a point taken arbitrarily in space, straight lines to their opposite faces, and if we take the plane passing through any three of the four points of intersection, then any line drawn from the intersection of this plane with the fourth face, to the line line dropped to that face, is cut by this straight line and by two faces of the tetrahedron in harmonic proportion.

Now, it is obvious that the theorem of (N.◦ 3) is but a particular case of this proposition. Moreover, the matter is of immense [30] wealth; witness, among others, the work of *Naudé*,[52] which is entitled: *Trigonoscopiœ, etc.*,[53] published in the *Mélanges de Berlin.* This Geometer comparing the theorems given above (N.◦ˢ 6.ᵉ, 7.ᵉ, 8.ᵉ), and considering on each side of a triangle the three points assigned in these theorems, namely: the midpoint of the side, the point where it is cut by the line that divides the opposite angle into equal parts, and the point where it is meets the perpendicular dropped from the opposite angle, formed, according to the different combinations of these points, classes or families of problems, the number of which was 678, and from there, following the remark of *Kaestner*,[54] demonstrated, among other things, how far we are from having exhausted all there is to say about even triangles in the plane.[55]

[52] Philippe Naudé (1684–1745).

[53] This is reference [35] in the current work.

[54] Abraham Gotthelf Kästner (1719–1800).

[55] *Mathematische Anfangsgründe: Anfangsgründe der Arithmetik, Geometrie, ebenen und sphärischen Trigonometrie, und Perspectiv.*, reference [24] in the current work.

Chapter 3

Practice

[31] **First Problem.** To prolong a line beyond an obstacle that does not allow us to see the two points placed in the direction of the line.

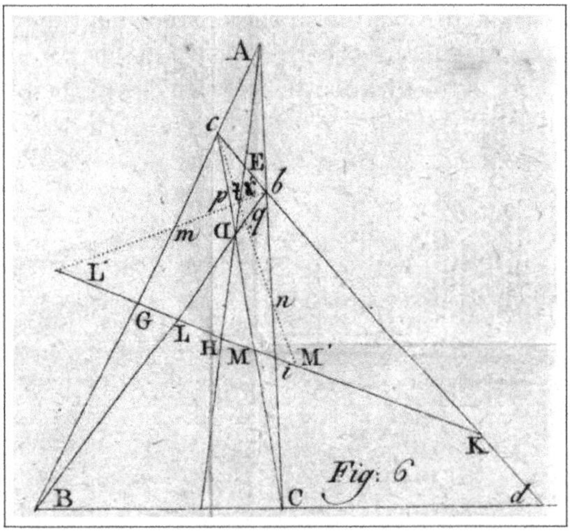

Figure 6. To prolong a line beyond an obstacle, First Solution

Fig. 6 19. **First Solution.** BC is the line that we seek to prolong.[1] From the point A, not on the line, and where we can see B and C, place a Stake. Place a second on AB at c, then a third on AC at b, such that the alignment[2] bc passes in front of the obstacle[3] O that hides the points B

[1]See [2, p. 128] for an additional diagram for this construction. Also, see [2, p. 176, 184] for a classroom activity for this passage.

[2]See p. 1, footnote 2.

[3]The obstacle O is not presented in Figure 6, because its position can change. This is explained by Servois below. However, one should envision the point O between the points C and d in figure 26.

and C. Place a fourth Stake at D, at the intersection of the two alignments bB and cC. Place a fifth Stake on AD at t, a sixth on bt and Ac at m, then a seventh on ct and Ab at n. Finally, place an eighth Stake at the intersection of the two line segments bc and mn at d. This last Stake will be on the prolongation of BC.

This is an immediate application of what has been demonstrated (N.° 14), and we see that we might have taken, in place of the points m and n, the points p and q, the alignments of which with those of the points b and c, would determine the point d equally well.

[32] To find the actual prolongation of BC, we find by the same procedure, a second point in the direction of BC. We can even, to ensure the accuracy of the operation, find a third, a fourth, etc., which must all be in the same alignment.

If the obstacle is between the points seen from the point A – if, for example, we see the points B and d, when the first Stake is still at A, we place the second at will at c, the third at m, both on AB, the fourth at b on cd, the fifth at n on md and Ab, the sixth at t on bm and cn, and the seventh at D on At and Bb. Finally the eigth at C the point of intersection of cd and Ab, and this point C is on the line Bd.

20. **Second Solution.** Ad is the line that we seek to prolong. [4] Off this line, we place the first Stake at B, a second at c on BA, a third at C on Bd – Cc being the line on which the point of prolongation will be taken – a fourth at b on the two alignments cd and AC, a fifth at D, the intersection of the two alignments Bb and Cc, a sixth at a, on the two alignments Bd and AD, and a seventh at [33] f the point of intersection of AC and ac. Finally, an eighth at E, the point of intersection of Bf and Cc. This last Stake will be on the prolongation of Ad. Fig. 12

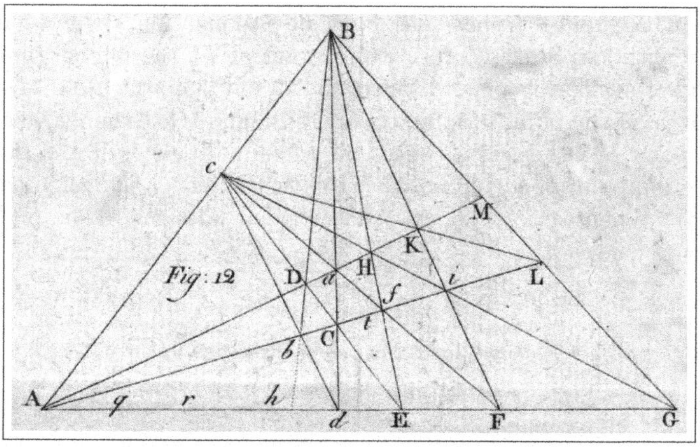

Figure 12. To prolong a line beyond an obstacle, Second Solution

[4] See [2, p. 130] for an additional diagram of this construction.

To have a second Stake on the same prolongation, I place a ninth at H on BE and aA, and a tenth at i on cH and Af. Finally an eleventh at F, the point of intersection of Bi and cf, the point F will still be on Ad.

If I had placed a twelfth Stake at K on BF and Aa and a thirteenth on cK and Ai at L, and finally a fourteenth at G, the point of intersection of BL and ci, the point G would also be on Ad. We see that one could continue indefinitely in this manner, as long as the investment in Stakes can be done conveniently.

This procedure is demonstrated, because following the construction, the lines AD and AC are so placed with respect to Ad that (N.◦ 3) all the transversals at the point B are cut by three lines AD, AC, and Ad in harmonic proportion.[5] On the other hand, following the same construction on each transversal BE, BF, BG, etc., cut by AD and AC, the points E, F, G, etc., are determined by the lines cC, cf, and ci, in [34] such away that (N.◦ 3) the segments BE, Bf, and BH of BE; the segments BF, Bi, and BK of BF; the segments BG, BL, and BM of BG; etc., are also in harmonic proportion. Therefore, the points E, F, G, etc., are on Ad.

If we wish, after having determined the two alignments AD and AC, and having marked the point a, as has been described, we take the alignments BH, BK, BM, etc., at will, which are cut by AD and AC, namely: BH at H and f; BK at K and[6] i; BM at M and L, etc., then we place the points t, f, u, etc., at the intersections of AC with Hd, Kd, Md, etc. Finally, we will mark the points of intersection of at and BH, for af and BK, for au and BM, etc., namely: the points E, F, G, etc., which are all on the alignment Ad. This is a modification of the Solution that may offer several advantages, for finding a sequence of points on the prolongation of Ad. This is demonstrated by the properties remarked (N.◦ 14).

Figures 9, 10 21. **Third Solution.** PQ is the line to be prolonged and bc is the line on which I wish to have the point of prolongation.[7] Having marked b and c with two Stakes, I fix a third point at D, the intersection of bP and CQ. I arbitrarily place a fourth at A, and [35] a fifth on AD at t, a sixth at x on the point of intersection of Ac and tQ, then a seventh at u at the point of intersection of Ab and Pt. Finally an eighth at R at the intersection point of bc and ux, and this last one is on the alignment PQ (N.◦ 16). When we know how to find a point on a line to be prolonged, we may find several.

---

[5]See p. 8, footnote 8.

[6]In [43, p. 34], this was given as f.

[7]See [2, p. 133] for an additional diagram of this construction.

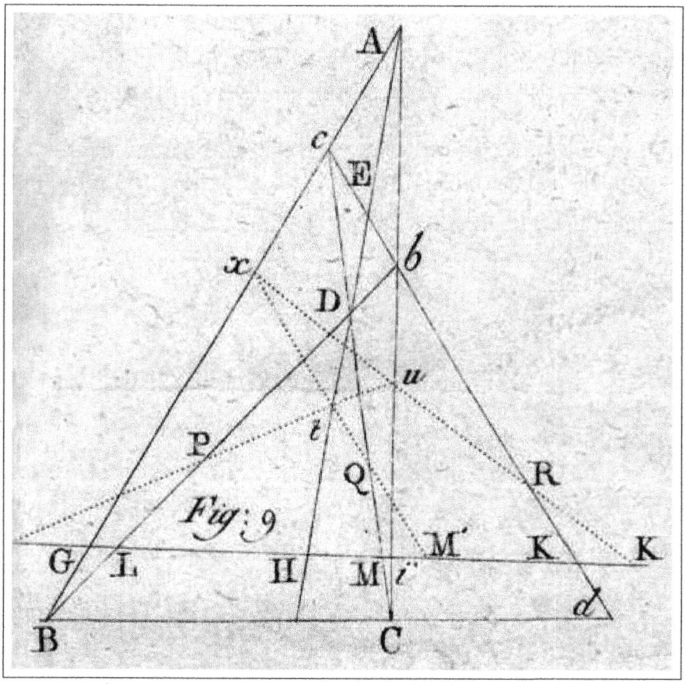

Figure 9. To prolong a line beyond an obstacle, Third Solution

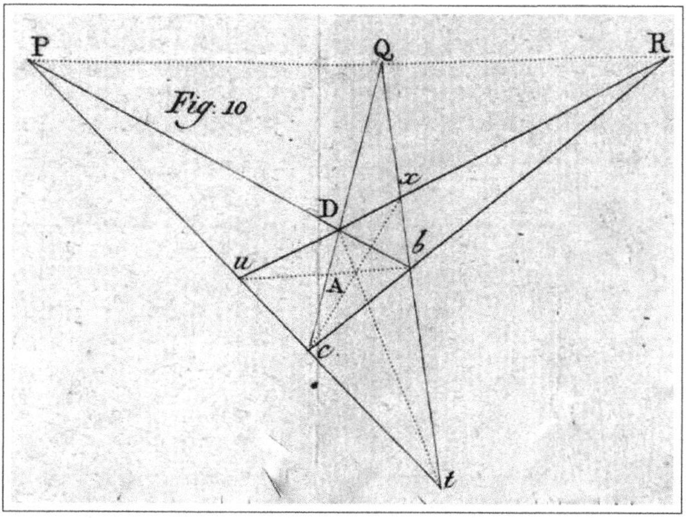

Figure 10. To prolong a line beyond an obstacle, Third Solution

Figures 7, 8

22. **Fourth Solution.**[8] PR is the line that we must prolong. From the points A and b from which I can see the points P and R, I place two Stakes, I fix a third at D, at will, in the alignment of bR, a fourth, also at will, at c on AP, a fifth at t on bP and AD, and a sixth at x on bc and AR. Finally a seventh at Q, the intersection of the alignments tx and Dc, and this point Q will be on the line PR (N.◦ 15). By the same procedure, we find a second point on the same line, and it is prolonged.

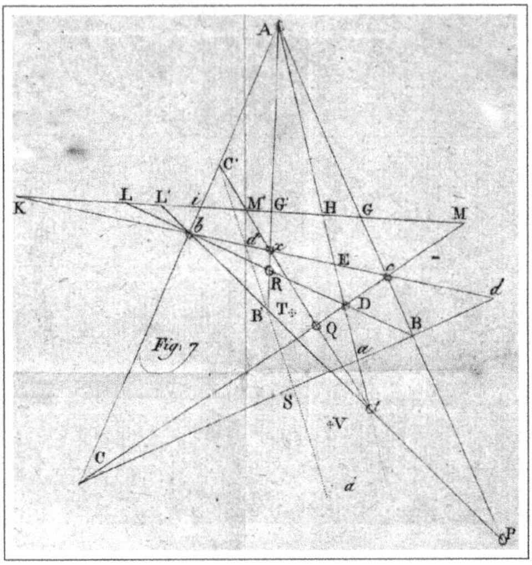

Figure 7. To prolong a line beyond an obstacle, Fourth Solution

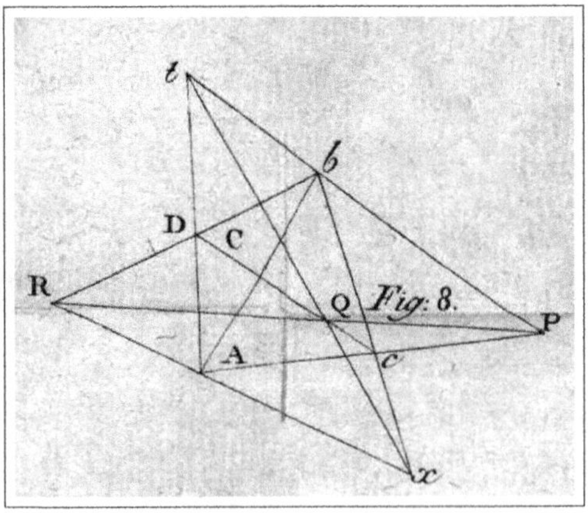

Figure 8.a. To prolong a line beyond an obstacle, Fourth Solution

[8]See [2, p. 134] for an additional diagram for this construction. Also, see [2, p. 177, pp. 178–82] for a classroom activity for this passage.

I could add other Solutions if it were necessary, because every time I have given three points on a straight line in the preceding theory, I have advance the principle of a particular solution to the problem. [36]

23. **First Observation.** The points A, B, and C (first Solution) may be inaccessible.[9] It is sufficient for this that we know how to place a Stake between two inaccessible points and on their alignment, but this is an extremely simple operation. Indeed, you have two indicators walk towards the line carrying Stakes, looking at each other and plant their Stakes simultaneously when they realize the inaccessible point in alignment towards which they are facing, of the Stake of the other indicator and the one in ones own hands.

Second Observation. Let use remark on this occasion that a single indicator can easily place a Stake at the point of intersection of two lines, each marked by two Stakes, because coming close to this point he will retreat or advance until he has discovered the point where the four visual rays pointing towards the four Stakes determining both lines, are reduced to two. This point is the intersection of the two line.

Third Observation. When it is a matter of placing a Stake at the intersection of two diagonal lines of a quadrilateral, which are determined by four Stakes, an indicator may be guided by the commands of the two observers [37] placed at the extremities of these diagonals. However, it is easier to prolong one side of each diagonal by an auxiliary Stake, and then this case reduces to the previous one, and a single indicator may mark the intersection of the diagonals.

Fourth Observation. In the second Solution, the three points A, B, and d may be inaccessible. In the third one, these are the points A, P, and Q. In the fourth one, the four points A, b, P, and R may all be inaccessible.

Fifth Observation. In the first Solution, if the alignments Ac, bD, Ab, and cD are given in any manner whatsoever, the points B and C could be invisible.

In the third Solution, if the alignments bD, cD, tx, and tu are given, the points P and Q may also be invisible.

In the fourth Solution, if the alignments bR, bP, AR, and AP are given, the points P and R may also be invisible.

In the second Solution, we must see the point A to determine the alignment AD; however, if the directions BC and bc are given, it will no longer be necessary to see the point d. Furthermore, if one wants to make use of the solution of the next problem to draw DA at the point of intersection of the given BA [38] and CA, one might also happen to see the point A.

---

[9] See [2, p. 135] for a diagram to aid in the understanding of the following observations.

Thus, we may always solve this particular problem:

> Given the directions of two lines for which the point of intersection cannot be seen along with the directions of the other two lines in the same case as the first ones, to find a point that is on the alignment of these invisible points of intersection.

Sixth Observation. In the operations detailed above, like those we have yet to explain, we must be careful to place the Stakes vertically, so that they are all parallel, this facilitates the alignments and makes them more exact. Also, because our theory, which supposes that everything lies in a single plane, which for simplicity we take to be horizontal, it is necessary that the points of the Stakes to be taken to meet this plane, which requires that they be perpendicular to it.

Seventh Observation. It is necessary to give these constructions the greatest generality as we can, because the more the Stakes that determine these lines that are removed from each other the easier it is to place new ones precisely in their [39] alignment. The arbitrariness that enters into the placement of the first Stake allows us to make the best use of this principle on the terrain that we are operating on. It is also necessary to avoid very acute angles as much as possible – it is difficult to accurately mark them – nevertheless, if you obliged to mark the point of intersection of two lines forming a very acute angle, we must not overlook the following correction: Having first fixed a point where the four visual rays drawn from the Stakes that determine the lines begin to reduce to two, we retreat or else advance to a second point where the two rays begin to divide into four, then place a Stake at in the middle of the interval between these two points. It will be close enough to the point of intersection of the two lines.

Eighth Observation. The problem has been solved only with Stakes, without measuring any lines or any angles. Mascheroni, in the Small Work cited (Book 2, Problem I) gives eleven Solutions to the same problem, but they are, for the most part, trigonometric, and all require the measurements for all the lines.

Ninth Observation. I have insisted on this problem, because it is used frequently in war, [40] and that it is particularly helpful to the Artillery Man, whether in the field where there is no better way than to fire in a row or in a siege, where it is necessary to fight the ricochet off the faces of fortifications. In this latter case, the lack of relief on the fortifications or other obstacles prevent us from immediately making prolongations; however, the enemy cannot help showing the muzzle of the pieces of which it arms its fortifications from the view of the observer field of fire. These are the points that must be used to apply our Solutions, which we might prefer by those that are provided by the instruments we know, even if we have them on hand, because they all require the measurements of lines that are totally unavailable to us. It could nevertheless happen that the face that we wish to prolong doesn't present any distinct points, if, for example, the cannons on which it is armed only fire with a ricochet, etc., if we can observe two furrows in the field traced out by a projectile fired from the fortification, we will have the direction of a line passing through the cannon. Additionally, if for the same piece we have two of these two different directions, and the same given for a second piece, we will have everything we need [41] to get the prolongation we wish to find (see the Fifth Observation above).

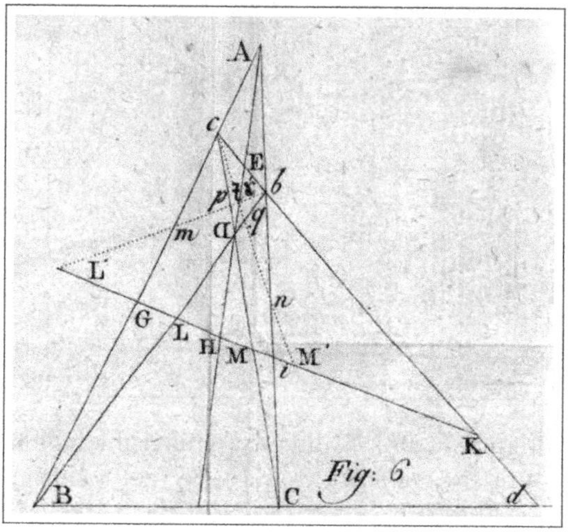

Figure 6. Draw a line to the invisible point of intersection of two lines, First Solution

24. **Second Problem.** From a given point to draw a line to the point of intersection, which cannot be seen, of two lines whose directions are given.

First Solution. Both lines BC and bc are marked by four Stakes, Fig. 6
and m is the assigned point given by a fifth Stake.[10] I assume the three Stakes B, c, and m are in the same alignment. If they were not, it would be easy to put them in alignment. I place a sixth Stake at the point of intersection A of the lines Bc and bC, a seventh at D, the intersection of the diagonals Bb and Cc; and then an eighth at t, the intersection of bm and AD. Finally a ninth at n, the intersection of Ab and ct, and the line mn is the one that is sought, because it intersects with BC and bc at the point d (N.◦ 14).

Whether the point m is placed between the lines BC and bc, or not, it is clear that the construction is still the same.

[10]See [2, p. 136] for an additional diagram of this construction and [2, pp. 178–82] for a classroom activity for this passage.

Figures 7, 8 **Second Solution.** *Ac* and *bt* are the two given lines, marked by four Stakes, and *R* is the assigned point mark by the fifth.[11] I place a sixth Stake at *D* on *bR* and *At*, a seventh at *x* on *AR* and *bc*, then an eighth at *Q* on *tx* and [42] *cD*, and *QR* is the required direction. Indeed, the two lines *bt* and *Ac* intersect at *P* (N.◦ 13).

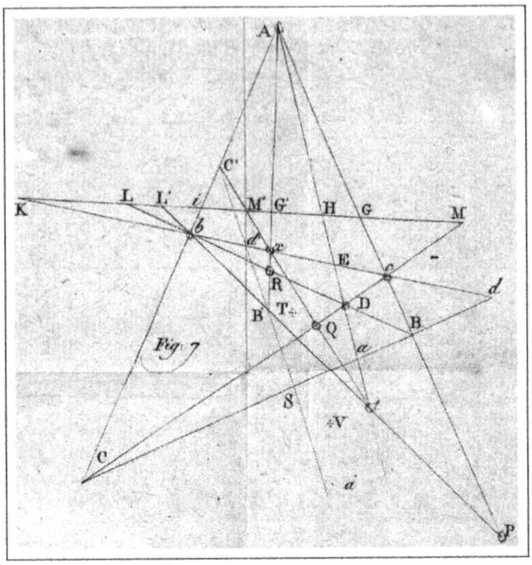

Figure 7. Draw a line to the invisible point of intersection of two lines, Second Solution

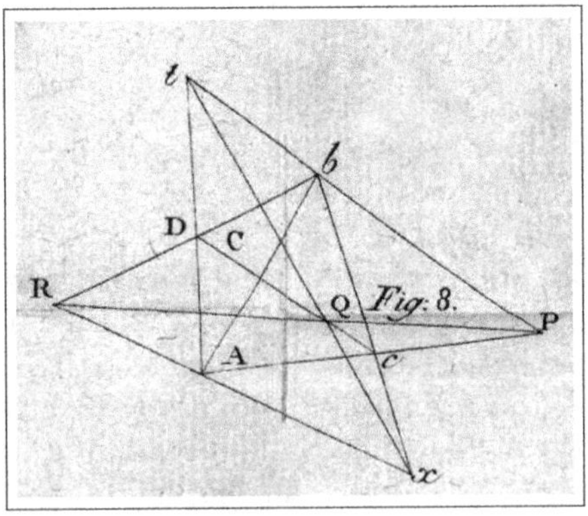

Figure 8.a. Draw a line to the invisible point of intersection of two lines, Second Solution

[11]In [43, p. 41], this Second Solution was given as N.◦ 25; however, that was in error based on Servois' formatting of item numbers. See [2, p. 137] for an additional diagram for this construction.

There is no need to add other solutions for this same problem. It is solved in many treaties of elementary Geometry, by two parallel lines and measured lines, or else by means of trigonometry.

25. **Third Problem.** From a given point, to draw a line parallel to two other parallels given in direction.

The Solution is the same as the proceeding problem, because the parallels are considered to intersect at an infinite distance.

26. **Fourth Problem.** To draw two parallel lines on the ground.

The solution is always possible in the following ways.

First Solution.

Two alignments on the same but very distant object, give two lines as parallel, which can be determined, when the object is at a great distance.

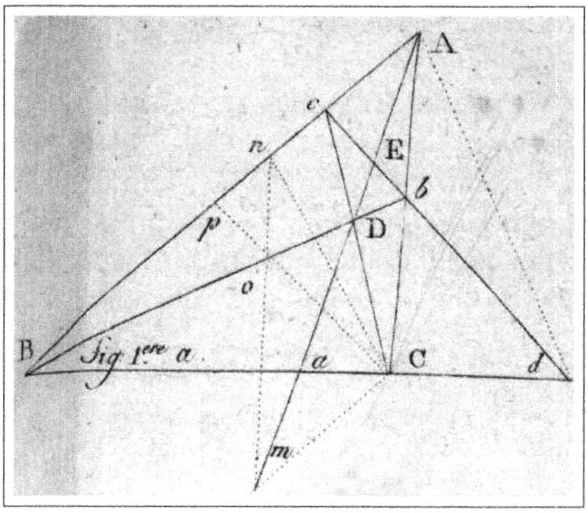

Figure 1.a. To draw two parallel lines on the ground, Third Solution, and draw a parallel to an accessible line, First Solution

Second Solution. The solar or lunar shadows of two Stakes observed at the same time, are two parallel lines. Two alignments taken at the same time on one star, are two parallel lines.

Third Solution. BC being any line to which we wish to draw parallel, I take [43] any length Ba, which I carry from a and C, and towards an arbitrary point A, I draw three lines BA, CA, and aA. I take any point b on CA, and I draw Bb which cuts aA at D. Finally I draw CD that cuts BA at c, and the line bc will be parallel to BC. Fig. 1.a

Indeed, we have, by formula (B) (N.◦ 3), because $Ba = Ca$ and $Bd = Cd$, which cannot be true unless the point d that meets bc and BC, is at an infinite distance, that is to say unless bc is parallel to BC.

27. **Fifth Problem.** From a given point, to draw a parallel to an accessible line.

Fig. 1.a **First Solution.**

The Third Solution to the proceeding problem immediately solves this one, because the point b, which we have taken arbitrarily on AC, will here be the given the point. Thus, BC being the inaccessible line, I draw through b any line bC that cuts BC at C. From that point, I take two equal lengths Ca and aB, I draw Bb that cuts aA at D, finally I draw CD to cut AB at c, and bc is the required parallel.

Second Solution. We will see at the end of the next problem, how we can draw from the considerations of the first and second solutions [44] (N.◦ 26), the way to solve the current problem.

28. **Sixth Problem.** From a given point, to draw a parallel to an inaccessible line, of which we see two points.

Fig. 13 **First Solution.**

xz is the inaccessible line of which we see the points x and z. A is the given point. I place at A the first Stake, place the second at B in the direction Az, then a third at C, at will. Through the point A, I draw a parallel to the accessible line Cz, by the procedure described in the proceeding problem. This parallel will cut BC at D. Through the point D, I draw a parallel to the accessible Cx, which cuts Bx at E, AE being the required parallel.

Indeed, $zBCx$ and $ABDE$ are two quadrilaterals with a common angle B, the sides DE and Cx parallel, and thus so are the diagonals AD and Cz. Therefore, the sides AE and xz are parallel.

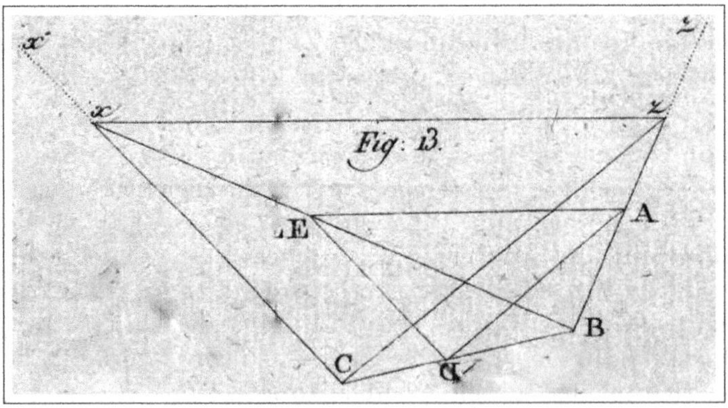

Figure 13. Draw a parallel to an inaccessible line, First Solution

29. **Second Solution.** The point C that we took at will (first solution) can be chosen in such a manner that the line Cz and Cx prolonged, pass through two very distant objects, z' and x'. Then we will mark the

point D on BC in the alignment Az' and the point [45] E on Bx in the alignment Dx', and AE will again be the parallel that was sought.

30. **Third Solution.** If we have any parallel to Cx, and any other parallel to Cz, we may determine the parallels AD and DE by the procedure of the third problem.

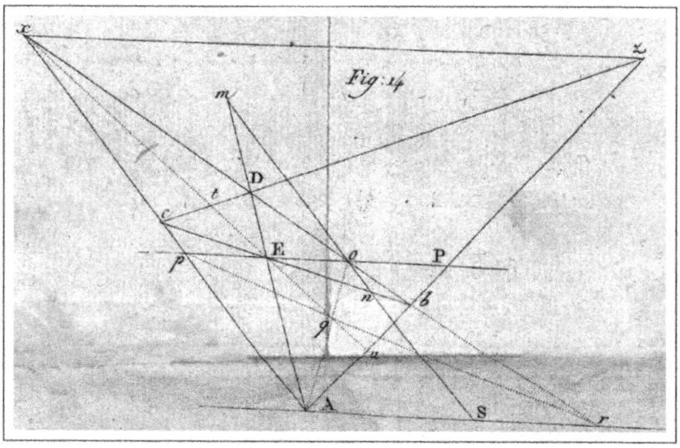

Figure 14. Draw a parallel to an inaccessible line, Fourth Solution

It is clear that the last two Solutions are *a fortiori* applicable in the case of an inaccessible line. This is in relation to the promise made above (N.◦27, Second Solution).

31. **Fourth Solution.** xz is the inaccessible line, and A the point Fig. 14
given.[12] I take the points b and c at will on the alignments Az and Ax; I mark the point[13] D, the intersection of bx and cz, and the point E, the intersection of the two diagonals AD and bc. I take the length AE from E to m, and the length cE from E to n. I draw mn that cuts bx at O, and EO is parallel to xz. I take the length om from o to S. I draw AS, and the problem is solved.

Indeed, by the construction, Am being the double of AE, and the double of cE, mn is parallel to Ac, and consequently po is the double of Ep, or $Ep = EO$. However, by the theorem (N.◦ 3), any transversal passing through E and intersecting xz at d, [46] gives the equation:

$$dp \times Eo = Ep \times do.$$

Therefore, when $Eo = Ep$, we have $dp = do$, which cannot happen unless in the case when Eo is parallel to xz. When mS being the double of mo, it is clear that As is parallel to Eo, and consequently to xz.

If the alignments Ax, bx, Az, and cz were given in any way, it is would not be necessary that we see the points x and z.

[12]See [2, p. 144] for an additional diagram for this construction.
[13]In [43, p. 45] this was given at the point d.

There is no need to take os equal to mo for determining the parallel AS. For this, is suffices to mark the point q the intersection of Ao and Ex, and the point r of the intersection of pq and bx, and Ar is the parallel (N.◦ 27). However, in this case, it is necessary that the point x be visible.

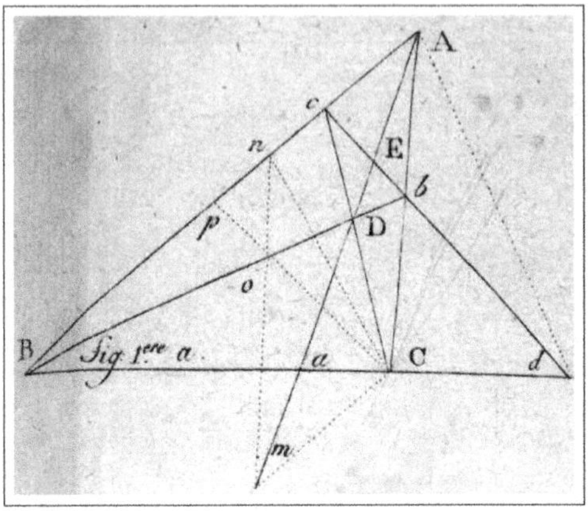

Figure 1.a. Divide a line into two equal parts, Solution

32. Mascheroni (Book 2, problem 3) gives eight Solutions to this problem. The first and the second do not suppose the use of any instrument; however, they require the measurements of four or five lines and the substitution of them in complicated formulas. The third, fourth, fifth, sixth, and seventh require calculation and the use of the Équerre d'Arpenteur.[14] The eighth uses the graphometer,[15] and a trigonometric calculation that is fairly complicated.

33. **Seventh Problem.** To divide a line into two equal parts.

Fig. 1.a **Solution.** [47] BC is the given line.[16] I draw the parallel bc to it, and I draw to any point A the lines BA and CA, which cut the parallel bc at b and c. I place the point D at the intersection of the two alignments Bb and Cc, and the line AD prolonged cuts BC at a into two equal parts. This Solution is the inverse of the first part of N.◦ 27.

We see that the extremities B and C of the line may be inaccessible.

[14]See p. xiv for more information about this device.

[15]This is a surveying tool used for angle measurements and sometimes called by surveyors as a semicircle.

[16]See [2, p. 145] for an additional diagram of this construction.

34. **Eighth Problem.** To divide a given line into any number of equal parts.

Solution. AG is the given line.[17] I draw AB, at will, and from any point[18] c on AB I draw cL parallel to AG, which cuts BG at L. I mark the point i, the intersection of AL and cG. I draw Bi that cuts AG at F, and we have, by the proceeding problem, $AF = \frac{1}{2}AG$. Fig. 12

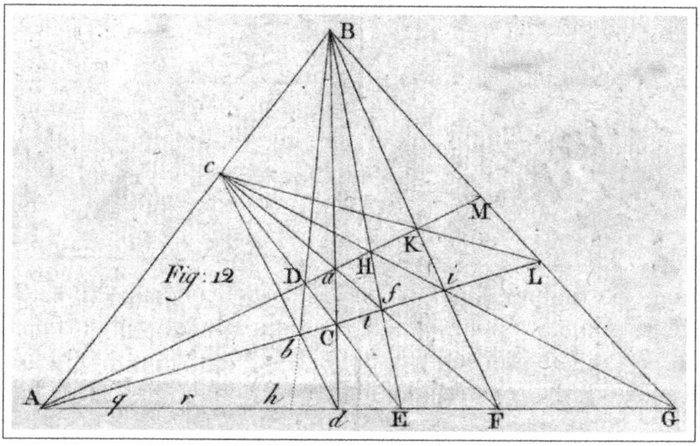

Figure 12. To divide a given line into any number of equal parts, Solution

I mark the point f, the intersection of cF and AL, and I draw Bf that cuts AG at E, and I have $AE = \frac{1}{3}AG$.

I mark the point C, the intersection of cE and AL, and I draw BC that cuts AG at d, and I have $Ad = \frac{1}{4}AG$.

I mark the point b, the intersection of cd [48] and AL, and I draw Bb that cuts AG at[19] h, and I have $Ah = \frac{1}{5}AG$.

We see the way we must continue to obtain a segment that is

$$\frac{1}{6}AG, \frac{1}{7}AG, \ldots \frac{1}{n}AG,$$

for n being any whole number.

Let[20] $Aq = \frac{1}{n}AG$. We will have

$$Gq = \left(\frac{n-1}{n}\right) AG.$$

I seek by the process that we have just indicated a final segment qr on Gq,

[17] See [2, p. 147] for an additional diagram of this construction.

[18] In [43, p. 47], this was given at the point e.

[19] In [43, p. 48], this point was given as H.

[20] In [43, p. 48], the line segment AG was missing in the following equation.

that is equal to

$$\left(\frac{1}{n-1}\right)Gq,$$

thus Gr will be

$$=\left(\frac{n-2}{n}\right)AG.$$

We will detach in the same way the final segment

$$\left(\frac{1}{n-2}\right)Gr,$$

and the remainder will be

$$\left(\frac{n-3}{n}\right)AG,$$

and so on. So that we finally come to the remainder $=\left(\frac{1}{n}\right)AG$, which will have its extremity point G, and consequently, we will be able to mark along AG all the divisions required.

Fig. 12 Indeed, from the construction and the theorem (N.◦ 3), we have

$$AE=\frac{AG\cdot AF}{AG+FG}, \quad [49] \quad Ad=\frac{AF\cdot AE}{AF+EF}, \quad AK=\frac{AE\cdot Ad}{AE+Ed},$$

and so on. However, $AF=FG=\frac{1}{2}AG$. Therefore[21]

$$AE=\frac{\frac{1}{2}\overline{AG}^2}{\frac{3}{2}AG}=\frac{1}{3}AG.$$

Thus,

$$EF=\frac{1}{2}AG-\frac{1}{3}AG=\frac{1}{6}AG.$$

Therefore, $Ad=\frac{1}{4}AG$. Thus,

$$Ed=\frac{1}{3}AG-\frac{1}{4}AG=\frac{1}{12}AG.$$

Therefore, $AK=\frac{1}{5}AG$, and so on. In general, we see that the sequence of segments $AE, Ad, AK, \ldots$ is recurrent, and that any term in the sequence depends on the two preceding terms, following the relation expressed by the equation

$$c=\frac{a\cdot b}{2a-b},$$

a, b, and c being any three consecutive terms.

[21] In [43, p. 49], Servois used an overline to indicate a quantity, similar to the way we would use a set of parentheses.

Let $b = \frac{1}{m}AG$ and $a = \frac{1}{m-1}AG$, we will have

$$c = \frac{\frac{1}{m(m-1)}\overline{AG}^2}{AG\left(\frac{2}{m-1} - \frac{1}{m}\right)} = \frac{1}{m+1}AG,$$

but we have seen that the stated law [50] will be true for the first integer values of m, therefore it is true for all of them.

This problem could be used to form on the terrain a scale of elevation, if we do not have one of the measures of steps at our disposal but rather we knew the total length of the line taken for the scale.

35. **Ninth Problem.** To divide an accessible angle into two equal parts.

Solution. From the vertex A of the given angle xAz, I take any two lengths AD and DB, on the side Az, I then take on the other side $AE = AD$ and $EC = DB$.[22] I mark the point F of the intersection of BE and CD, and the line AF divides the given angle in to two equal parts. This is clear. Fig. 15

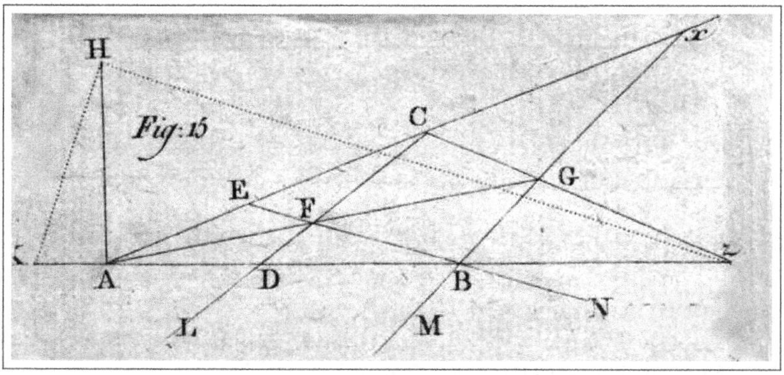

Figure 15. To divide an accessible angle into two equal parts, Solution

[22]See [2, p. 150] for an additional diagram of this construction. Also, see [2, p. 187] for a classroom activity for this passage.

36. **Tenth Problem.**[23] To divide an inaccessible angle into two equal parts.

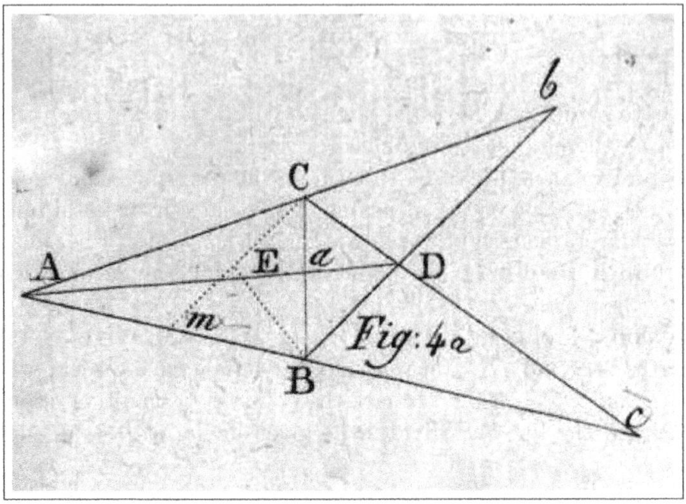

Figure 4.a. To divide an inaccessible angle into two equal parts, First Solution

Fig. 4.a **First Solution.** BAC is the inaccessible angle.[24] B and C are accessible points taken on the sides of the angle. I join BC. By the proceeding problem I divide the angles ABC and ACB each into two equal parts, and the point of intersection E of the lines that divides them is one of the points on the line that splits the given angle into two equal parts. This is a consequence of the noted theorem (N.◦ 8).

If we can see the point A, the direction AE will be that of the line that divides the given angle into two equal parts.

If we cannot see the point A, we will make use of the second problem, to draw, through the point E, a line that intersects with the lines in the given directions BA and CA. Or else we will take on AB and AC other points on which we will repeat the same construction to obtain a second point on the line that divides the angle in to two equal parts.

This Solution solves the problem of finding the prolongation of the capital of a fortress[25] in a very simply way.[26] However, it is not very practical when the flanking angle is very acute. When the fortress is flat, the point E is too close to the salient.[27] This is why I add this other Solution.

[23]In [43, p. 50], this was given as "Problème deuxième," literally the Second Problem.

[24]See [2, p. 151] for an additional diagram of this construction.

[25]See p. xiv for more information about French fortification terminology.

[26]See [1, p. 424] for a detailed description of this problem. Also, see [2, pp. 188–90] for a classroom activity for this passage.

[27]This is a military term, which means the object pointing out.

37. **Second Solution.** Instead of dividing the angles ABC and ACB into two equal parts, I make this division on their supplements CBc and BCb by the lines that intersect at the point D. This point D will be the capital of the angle.[28] Fig. 4.a

It may happen that we cannot see the salient A, and we cannot place the point D, either because it is too far away, or because [52] it falls into a ravine, etc. However, we do have the directions BA, BD, CA, and CD. Therefore, we may (N.◦ 23, Fifth Solution.) find two points on the alignment of the invisible points of intersection A and D.

38. The solution[29] that we usually bring to this problem measures the angles ABC and ACB, then taking half of their sum that we construct at C, for example, in placing a side on CA, and prolonging the latter until it meets AB. By this method we have an isosceles triangle, whose middle point of the base is the point of the capital[30] of the angle. However, measuring angles and constructions on the ground cannot executed, without instruments, in a fairly accurate manner. Besides, we must chain the base, which might be quite long, one and half times, if the salient angle is obtuse, and if we do not want to get too close.

39. **Eleventh Problem.** To erect a perpendicular on a given line at a given point.

Solution. *ad* is the given line that on which we wish to raise a perpendicular at the point *a*.[31] I take any length *ab*, which I take to be at *c* in an arbitrary direction of *bc*. I take the length *ac*, from *a* [53] to *d* on *ad*, and from *c* to *E* on *bc*. I draw *df*, on which I twice take the length *ab* from *d* to *f*, and *af* is the perpendicular sought. Fig. 16

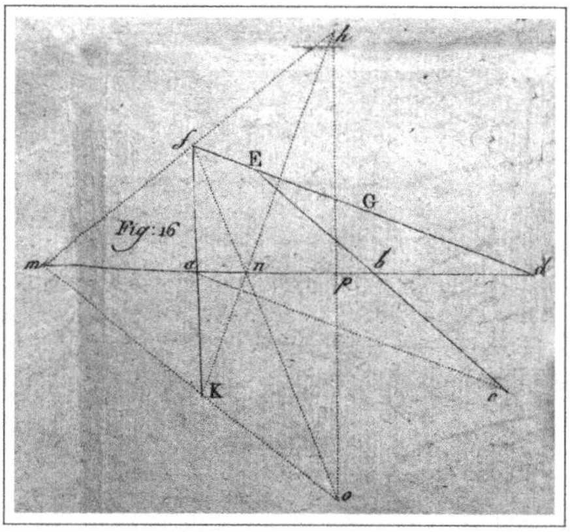

Figure 16. To erect a perpendicular on a line at a given point, Solution

[28] N.◦ 8 above. See also [2, pp. 152] for an additional diagram of this construction.

[29] See [2, p. 156] for a diagram to aid in understanding the solution.

[30] I.e., the bisector of the angle.

[31] See [2, pp. 157–58] for additional diagrams for this construction. Also, see [2, p. 191] for a classroom activity for this passage.

Indeed, if we draw the the line[32] aG through the point G, the midpoint of df, we see that the triangle aGd is isosceles. Because by the construction Ed is parallel to ac, and consequently the angle Gdb = the angle bac. The side $ad = ac$ and the side $dG = ab$. The two triangles aGd and abc are therefore equal, and the latter being isosceles, the first one is necessarily also. Therefore $aG = ab = fG$. The triangle afG is therefore also isosceles. The angle faG is consequently the half the exterior angle aGd and the angle baG is half of the angle aGf. Thus, the sum of the angles faG and baG, or the angle baf, is half of the sum of the two angles aGf and aGd, which together make up right angles. Therefore the angle baf is right.

Here is another solution of the same problem.[33]

Figures 4, b.

DE is the line on which we wish to erect a perpendicular. I take an arbitrary length[34] Do which I put twice on DE, that is: from D to o and from o to E, then in arbitrary directions, once from o to C and once from o to B. [54] I mark the point of intersection n of DC and EB, and the point of intersection p of BD and CE. This being so, the line np will be perpendicular to DE. Indeed, in the triangle DEn, the lines BD and CE are perpendiculars dropped from the two angles to the opposite sides. Therefore (N.◦ 6) np, drawn from the third angle to the point of intersection of the perpendiculars BD and CE, is also perpendicular to the third side DE.

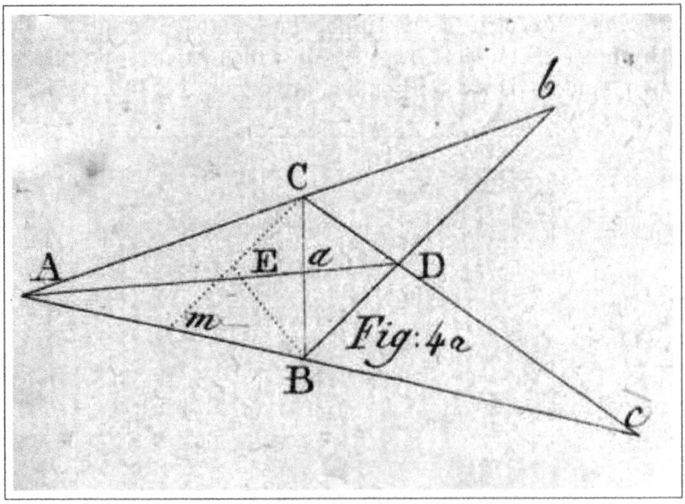

Figure 4.a. To erect a perpendicular on a line at a given point, Alternate Solution

[32]In [43, p. 53], the point G throughout this paragraph was given as g.

[33]See [2, pp. 159–60] for additional diagrams for this construction.

[34]In [43, p. 53], this length was given as do.

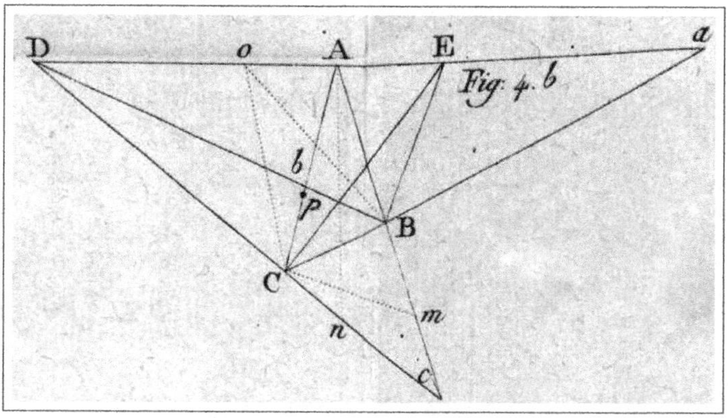

Figure 4.b. To erect a perpendicular on a line
at a given point, Alternate Solution

If we wish that the perpendicular is erected starting from the given point E, we will then take two equal lengths on the perpendicular An. I assume that these are the lengths Ap and pn, and we will be marked on En at the point B, where Dp meets it, and the point c on Dn, where it is cut by the direction AB. This being so, Ec is the required perpendicular, because by this construction Ec is parallel to An (N.∘ 27).

40. **Twelfth Problem.** From an inaccessible point, to drop a perpendicular to an accessible line.

Solution. ab is the accessible line, and h is the inaccessible point.[35] Fig. 16
From any point a on ab, I erect a perpendicular, by the procedure of the previous problem, it then a matter of drawing from this perpendicular [55] af a parallel passing through the point h. This will be the perpendicular sought. For this, if we wish, then proceeded as follows. I bring the length af, from a to k on af prolonged. I place on ab the point m, in the alignment fh and the point n, in the alignment kh. Finally, I mark the point o the intersection of fn and mk; and the line ho is the perpendicular required, which cuts ab at p.

[35]See [2, p. 161] for an additional diagram of this construction.

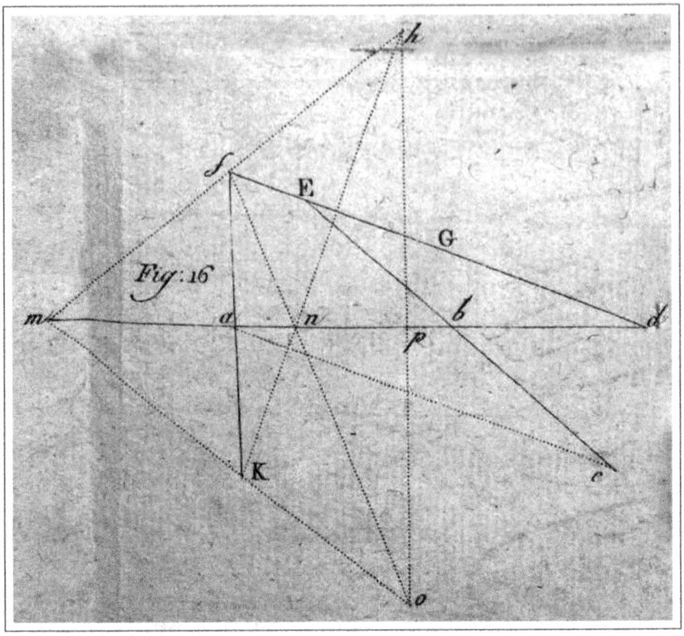

Figure 16. Drop a perpendicular to an accessible line, Solution

This construction puts the point o below ab, where the point h is above it. If the accidents of the terrain do not allow us to mark the point o, we might find the point p by one of the described procedures, (First Problem, N.◦ 23, Fifth Solution.), where the directions fn and mk may have as their point of intersection. We may also strengthen the construction by bringing it to the midpoint of af, and not to the midpoint of fk, etc.

It is necessary to note that this solution solves, a fortiori, the problem of a erecting a perpendicular from an accessible point to a given line.

41. **Thirteenth Problem.** From a given point, to drop a perpendicular to an inaccessible line of which one sees two points.

Fig. 14 **Solution.** xz is the inaccessible line of which we see the points x and z.[36] A is the point from which we must drop the perpendicular. By the procedure indicated (N.◦ 40), I drop the perpendicular bx from the inaccessible point x to the accessible line Az, and the the perpendicular cz from the other inaccessible point z to Ax. These two perpendiculars intersect at D, and AD is in the direction of the perpendicular drawn from the point A to xz (N.◦ 6).

[36]See [2, p. 162] for an additional diagram of this construction.

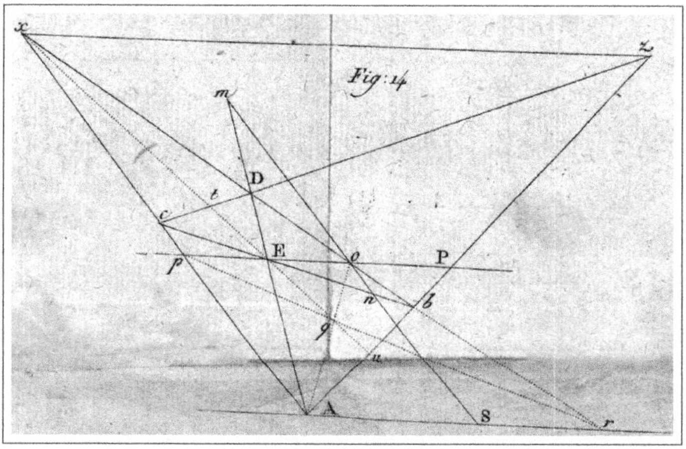

Figure 14. Drop a perpendicular to an inaccessible line, Solution

If you wish to have the distance from the inaccessible line, and further if we wish that this distance is from one of the points of the line that we can see (from x for example), then having, as described above, the direction of the perpendicular dropped from the point A, where we are, namely: AD, I erect a perpendicular on AD at the point A (N.◦ 39) – it will be parallel to xz – and on this parallel I drop a perpendicular from the inaccessible point x (N.◦ 40); this will be the distance sought. We will see below how to measure it.

42. Mascheroni (Book 2, problems IV, V, VI) occupies himself with the problems solved (N.◦ˢ 39, 40, 41). All of his methods require either rather long calculations or mathematical instruments.

[57] For the Solution to the problem (N.◦ 39), we know how to use a triangle made of cords whose sides are in the ratio of the numbers 3, 4, and 5. However, it is clear that they cannot give the right angle exactly, if the sides of the triangle are not very large, which will become very inconvenient. We must say the same of methods that assume arcs of a circle described in the field, because if the radii are small, the operation lacks accuracy, if they are very large, the operation is almost impossible.

43. **Fourteenth Problem.** To measure a line, of which one of the extremities is inaccessible.

Solution. Aa is the line we want to measure. Here the extremity A is inaccessible.[37] I place two Stakes at will at B and C, in the same alignment as the point a. I place another Stake at will on Aa at D. I mark b by a Stake at the point of intersection of BD and AC, and c by a Stake at the point of intersection of CD and AB. Finally E by the Stake at the intersection of bc and Aa. I measure the length Ea, and in noting, Figures 1, 2

[37] See [2, p. 163] for an additional diagram of this construction.

by the way, the lengths aD and DE, and I have the length of Aa by the equation

$$Aa = \frac{Ea \cdot Da}{Da - DE},$$

in the right hand side of which every quantity is known.

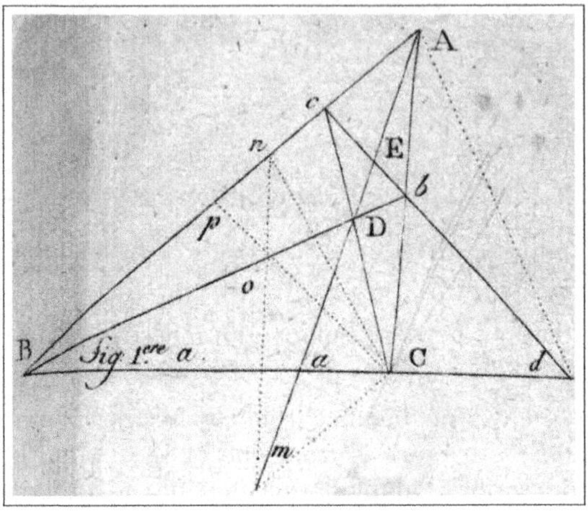

Figure 1.a. Measuring a line whose extremities are inaccessible, Solution

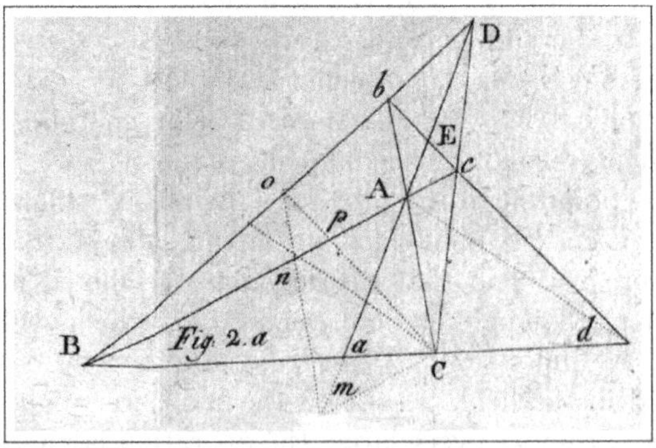

Figure 2.a. Measuring a line whose extremities are inaccessible, Solution

This is an immediate application of the theorem (N.◦ 3). It was positively indicated by Carnot (the correlation of the figures of *Géométrie*, N.◦ 191, page 135).

We may vary this construction greatly. For example, we can instead place the Stakes B and C on one side and the other of the point a, or both on the same side. The point D can be taken between the points a and A, or on one side or the other, but we will never be unable to find the formula into which we must substitute them, if we clearly understood the theorem (N.◦ 3).

We must note that one of the two points B and C may be inaccessible.

44. If we do not wish to make any calculations to solve the problem, we might use the following method. Az is the line that we wish to measure and whose extremity z is inaccessible.[38]. I note a direction AC on the ground, freely and at least of a length presumed to be a little bigger than Az. I place at will two Stakes at D and B on Az. On AC, I make $AE = AD$ and $EC = BD$. I place a Stake at F, the intersection of BE and CD, and I place another at G the intersection of AF and Cz. Finally, I determine the point x at the point of intersection [59] of AC and BG, and Ax is equal to Az. We can then can measure Ax with paces or with a chain. This construction does not need a demonstration. Fig. 15

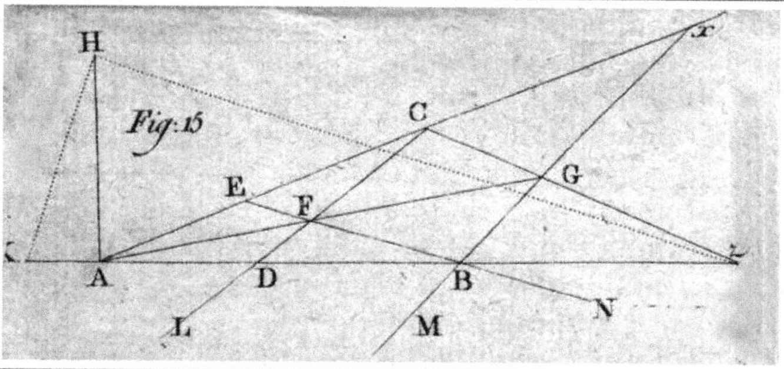

Figure 15. Measuring a line whose extremities are inaccessible, Alternate Solution

45. I pause here for a moment to compare the indicated Solution (N.◦ 43) with an analogous solution commonly proposed.[39]

We take at will the point D on Aa, the point C next to it, and on Ca the point B, and we determine on CD the point c in the alignment of AB. This being so, we have Fig. 1

$$Aa = \frac{BE \cdot Da \cdot Dc}{Ba \cdot CD - Ca \cdot Dc},$$

or else

$$Aa = \frac{BE \cdot Da \cdot Dc}{Ba \cdot Cc - BC \cdot Dc}.$$

This solution is given by Mascheroni (Book 1, problem II). He gives no demonstration. However, his formulas are simple modifications of those we have remarked (N.◦ 4).

By making specific assumptions, such as placing the point D at the midpoint of Cc or in general by making $Dc = m \cdot Cc$, m being any integer,

[38] See [2, p. 165] for an additional diagram for this construction.

[39] See [2, p. 166] for an additional diagram for this construction.

or else by supposing $BC = n \cdot Ba$, the formulas simplify and we encounter the Solutions [60] proposed by different Authors. However, in all cases we must *chain* the five lines Da, Ba, BC, Cc, and Dc.

We know the method attributed to Vauban.[40] It assumes the construction of right angles, or at least an angle equal to another, and requires a very long chaining.[41]

Remark. The current problem is of the utmost importance for the Artillery Men, who must always know the distance to the target, if he does not want to unnecessarily consume ammunition. The known solutions are sufficient when the target is fixed. However, they seem difficult to apply when it is moving, because given the time to execute them they cannot be repeated often enough to give the distance to the moving object at each instance. It is true that in this case, it is proposed to use two graphometers placed at the extremities of a common base; however this is a device that we will not always have, and it also requires the time to conclude from the two angles the distance that we need. However, it is singularly important that we may estimate at every instance, or at least at short intervals, the distance to an object, weather it is the [61] head of a column, a ship at sea, etc., which approaches or recedes. Then we will know when to start or cease firing. I may be permitted several points-of-views on this interesting matter; they are not a digression.

If the object that we must hit moves on the radius of a circle whose battery occupies the center, we may make good use of the *Triangle-équerre*[42] of *Lagrange*,[43] manufactured by *Delure* in *Paris* (this instrument made for Artillery Men should be known to them), for having measured ahead of time a base perpendicular to the direction of the moving target, terminating on one end of the battery, and fixing the the instrument at the other extremity, an observer, constantly holding the mobile alidade[44] directed towards the object, may state at every instance the distance from the battery to object, because this instrument gives the distance without calculation.

Fig. 1. c. d. E. If the object moves in a straight line, oblique with respect with the lines of fire of the battery, we will come to know its distance by proceeding as follows. AD is the direction of the movement of the target. Two observers simultaneously determine, near the battery, with Stakes the two [62] alignments CA and Ca directed to the object A that is out of range. They then determine two other alignments BD and[45] bD of the same

[40]Sébastien Le Prestre de Vauban (1633–1707). In [43, p. 60], the spelling was given as "Vaunban."

[41]See [55, pp. 37–43].

[42]This is an instrument used to measure a distance without directly comparing it with a measuring rod, which comes in useful when the measurement depends on angles. This device has been replaced by the modern device known as a "telemeter."

[43]Joseph-Louis Lagrange (1736–1813).

[44]This is a component of the Équerre d'Arpenteur.

[45]In [43, p. 62], this was given as cD.

object maybe having arrived at D, still out of range, these four alignments give the quadrilateral $BbCc$, and consequently the point of intersection d of the two diagonals, which is close to the battery, and we may, with little error, take this as the center. After this we place a Stake at d, from which two long cords are extended, divided meter by meter. An observer, holding one of these cords consistently taut, travels along BD so as to always be in the alignment with the Stake d and the object. Another observer, holding the other cord, travels along CD consistently in the alignment with the Stake d and the first observer. At small intervals marked by any signals, both observers tell the calculator their respective distances from the point d, and the observer has soon found the distance of the object to the same point, because it is equal to twice the product of the distances recorded divided by their difference. Because, if a' is the position of the object at any point of AD, and B' and C' being the corresponding [63] alignments of the observers on BD and CD, we have (N.◦ 3)

$$a'd = \frac{2 \cdot B'd \cdot C'd}{C'd - B'd}.$$

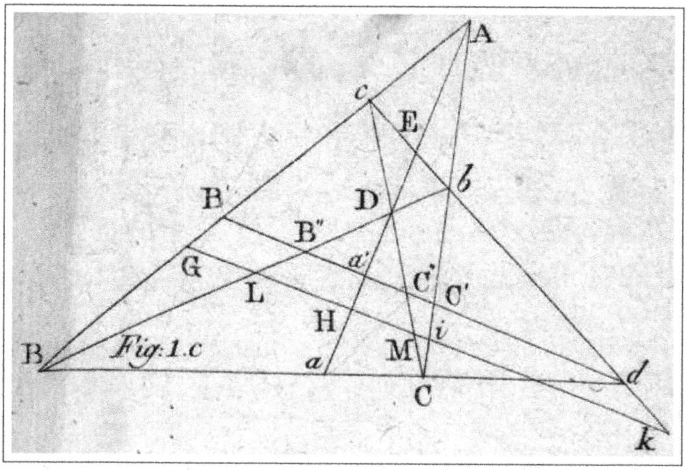

Figure 1.c. Measuring a line whose extremities are inaccessible, Mascheroni's Solution

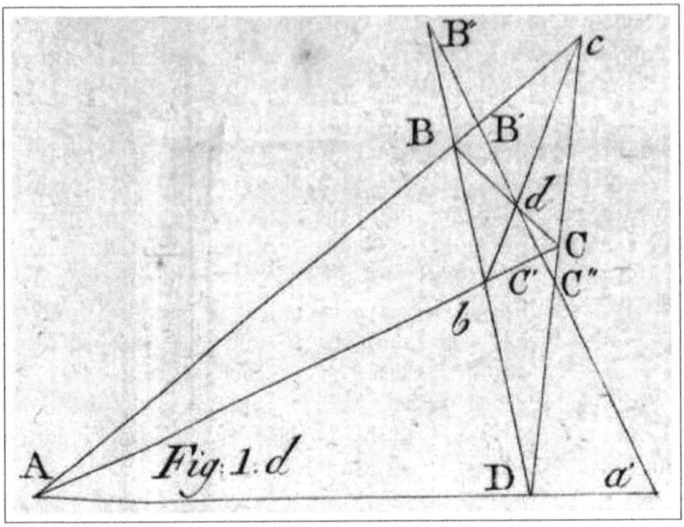

Figure 1.d. Measuring a line whose extremities are inaccessible, Mascheroni's Solution

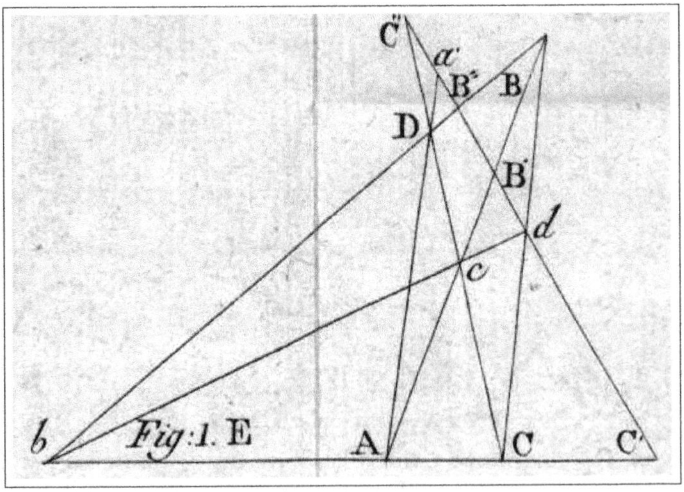

Figure 1.e. Measuring a line whose extremities are inaccessible, Mascheroni's Solution

If the target does not lie in a straight line, we may use the following method, which I have tried several times successfully.[46]

[46]In [43, p. 63], this is the only time that Servois makes a reference about his techniques being actively utilized.

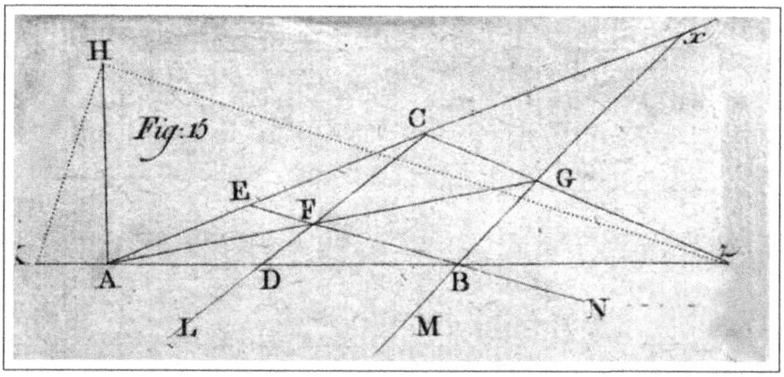

Figure 15. Measuring a line whose extremities are inaccessible, Mascheroni's Solution

A is the center of the battery, or close to it and *x* is the moving target. Fig. 15
Az is a base of length greater than the average range of the cannon of the battery. It is measured exactly, and, towards the extremity *z*, the divisions, meter by meter, are indicated by pickets, stones, etc. If the terrain allows it, the direction of this base is taken so that it does not form an obtuse angle with the first or last line of fire directed to the moving target. Stakes are placed at *A*, *D*, and *B* on *Az*, at arbitrary distances. A cord *AC* equal to *AB* and we on which we have marked at *E* a length equal to *AE* to *AD*, is fixed at one extremity at *A*. Eight men designated by *E*, *C*, *F*, *G*, *L*, *M*, *N*, and *z*, are given the following tasks. *C* and *E* hold the cord, one at *E*, the other at *C*, and are careful to be always in the alignment *Ax*. *L*, placed at a distance from the Stake *D*, stays constantly [64] on the alignment *CD*. *M* is always on the alignment *Bx*, and *N* is always on the alignment *BE*. *F* will constantly occupy the point of intersection of *DL* and *BN*, and *G* constantly always at the point of intersection of *AF* and *BM*. Finally, *z* is always at the intersection of *CG* and *AB*, and he proclaims at every instance at what distance he is from the point *A*. It is the same as that from the goal *x* (N.° 44). If the object moves in the direction *Ax*, it is clear that the observers *E*, *C*, *F*, and *M* may be replaced by Stakes and *L* and *N* become useless, and at *G* and *z*, with the attributions that have been given to them, can alone determine at every instant the distance *Az* equal to *Ax*.

46. **Fifteenth Problem.** To measure an inaccessible line.

Fig. 14 **First Solution.** xz is the inaccessible line. By the sixth problem, I draw a parallel Ar to them.[47] I measure, by the preceding problem, the distance Az, from which the extremity z is inaccessible. Finally, I measure the accessible lines Ab and Ar, and I have, as we know[48]

$$xz = \frac{Ar \cdot (Az - Ab)}{Ab}.$$

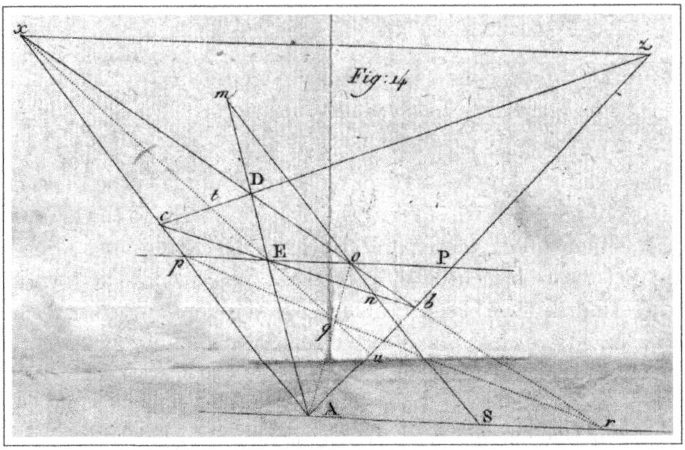

Figure 14. To measure an inaccessible line, First Solution

We may, if we wish, unite the two operations [65] as follows. I determine Eo as above (N.◦ 31), and I measure EP. I draw Ex that cuts cz at t, and Az at u, and I measure Et and Eu, and consequently tu. Given this, I have

$$xz = \frac{EP \cdot tu}{Eu - Et}.$$

Indeed, it is clear that $xz : EP :: ux : Eu$ or

$$xz = \frac{EP \cdot ux}{Eu}.$$

However, (N.◦ 43) we have

$$ux = \frac{tu \cdot Eu}{Eu - Et}.$$

Therefore,

$$xz = \frac{EP \cdot tu}{Eu - Et}.$$

[47]See [2, p. 168] for an additional diagram for this construction.
[48]In [43, p. 64], this was given as Xz.

47. **Second Solution.** xz is the inaccessible line.[49] I place a Stake at A, from which I can see x and z. Then another on Ax at B, a third on Az at C, a fourth at G, the intersection of Bz and Cx, a fifth at D, the intersection of AG and BC, and a sixth at F, the intersection of Ax and Dz. Finally a seventh at E, the intersection of Az and Dx. I measure the perimeter of the triangle ABC, and in distinguishing the parts AF, FB, BC, CE, and AE, given this, we have xz by the equation: Fig. 17

$$xz = \left(\overline{AB}^2 \cdot m\,(m-n) + \overline{AC}^2 \cdot n\,(n-m) + \overline{BC}^2 \cdot m \cdot n\right)^{\frac{1}{2}},$$

in which m and n [65] are known numbers: namely:

$$m = \frac{AF}{AF - BF} \quad \text{and} \quad n = \frac{AE}{AE - CE}.$$

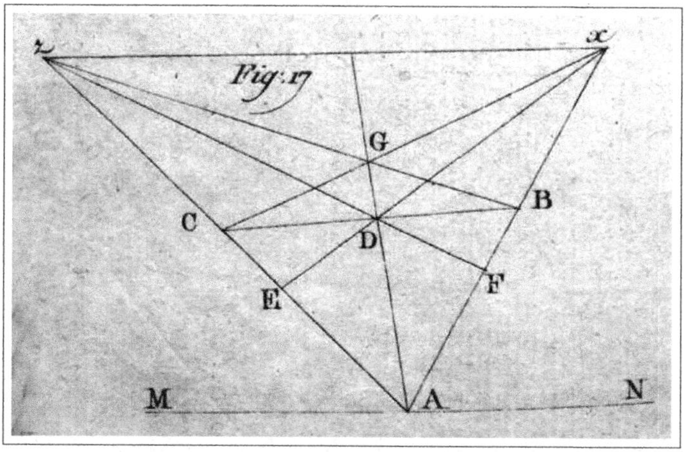

Figure 17. To measure an inaccessible line, Second Solution

Indeed, the two triangles ABC and Axz having a common angle A, it easy to see that by making $Ax = m \cdot AB$ and $Az = n \cdot AC$, we have the above formula, moreover, by (N.◦ 43), we have

$$Ax = \frac{AB \cdot AF}{AF - BF} \quad \text{and} \quad Az = \frac{AC \cdot AE}{AE - CE},$$

and consequently m and n have the stated values. If $m = n$, or if BC is parallel to xz, the formula gives $xz = m \cdot BC$, as above.

48. Mascheroni gives twenty Solutions to this problem (Book I, problem 3). The first eight do not require the use of instruments, they reduce

[49] See [2, p. 170] for an additional diagram for this construction.

for all intensive purposes, to our first. However, they also depend on the way of drawing a parallel to an accessible line, and of measuring a line for which one extremity is inaccessible. The others are provided with the aid of mathematical instruments.

49. **Sixteenth Problem.** To measure the distance from a point to an inaccessible line.

Solution. [67] First we must find, by the thirteenth problem (N.◦ 41), the position of this distance, that is the direction of the perpendicular from the point to the line.[50]

Fig. 12 AE is the inaccessible line of which we see the points A and E. BC is the direction of the perpendicular obtained by whatever method. I arbitrarily take the point c on AB. I mark the point of intersection of BC and Ec by C, then the point of intersection of BE and AC by f. Finally, the point a, the intersection of BE and cf, and and the length of the perpendicular to the line, namely, Bd, is found by the equation

$$Bd = \frac{Ba \cdot BC}{Ba - Ca}.$$

This is another application of theorem (N.◦ 3), and the problem (N.◦ 43).

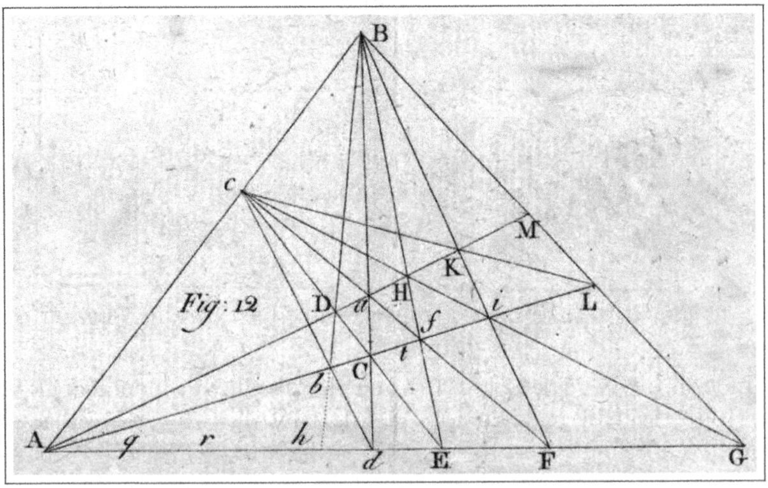

Figure 12. To measure the distance from a point to an inaccessible line, Solution

[50]See [2, p. 171] for an additional diagram for this construction.

50. **Alternate Solution.** After having executed the specified process (N.° 47), to measure the line xz, I know the three sides of the triangle Axz. This being so, $xz = a$, $Ax = b$, $Az = c$, and $a + b + c = 2p$. Denoting by P the perpendicular erected at A on xz, we have, as is known:[51] Fig. 17

$$P = \frac{2}{a}\left(p \cdot (p-a) \cdot (p-b) \cdot (p-c)\right)^{\frac{1}{2}}.$$

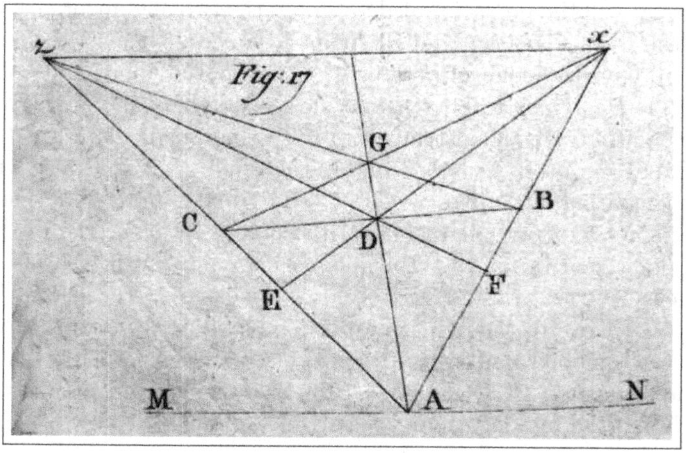

Figure 17. To measure the distance from a point to an inaccessible line, Alternate Solution

This Solution is faithful, for the most part, to what is commonly prescribed; however, it is long and laborious, and I doubt that we would be tempted to prefer it to the first one.

Little Known Uses of the Équerre d'Arpenteur.

51. Everyone knows the form of the Équerre d'Arpenteur,[52] so I will dispose with making a description of it. Moreover, its construction is varied at will. I prefer those that make up the Équerre[53] from two mobile alidades around a center, which we can set at right angles, when we wish, because the instrument gives in addition to the Équerre with fixed branches, the way to take an angle on the battle field and transporting it elsewhere.

[51]This is Heron's formula after Heron of Alexandria (c. 10–c. 70 CE).

[52]See p. xiv for more information about this device.

[53]This is a variation of the Équerre d'Arpenteur that can be used to draw vertical or diagonal lines, and verify and draw right angles. Again, the French term will be used.

Everyone also knows, how by means of the Équerre, to erect a perpendicular on a given line, to drop a perpendicular from a given point, even an inaccessible one, to drop a perpendicular on an accessible line, to draw a parallel through a given point to an accessible line, etc. So I will refrain from any details in this regard. However, I will show what help the Équerre can be in the Solutions to problems mentioned above. [69]

52. Solve the first problem with the Équerre.

Fig. 11 *Bc* is the line that we wish to prolong, of which we see the points *B* and *c*. I take at will the point *d* that determines the two alignments *Bd* and *cd*. I take the point *C* on *Bd*, and the point *b* on *cd*, *Cb* being the line which I want to have the point of prolongation. I take the point *E* at will, next to the side *Bd*, which will determine the second alignment at the point *B*. At the point *d*, with the instrument, I raise a perpendicular on *Bd*, which cuts *EB* at *m*, and I measure *dm*. At the same point *d*, I raise with the instrument a perpendicular *dn* on *cd*, and I make $dn = dm$. The point *n* determines a second alignment on the point *c*. I move the instrument to the point *C*, I raise the perpendicular *Cr* on *Bd*, which meets *EB* at *r*. I measure *Cr*. I raise a second perpendicular *Co* on *Cb*, and I make $Co = Cr$. I move to the point *b*. I raise, on *cd* a perpendicular *bq*, which cuts *cn* at the point *q*. I measure *bq*. I raise another perpendicular *bp* on *Cb*, and I make $bp = bq$. Finally, I draw *op*, which will intersect *Cb* at point *A* and the prolongation of *Bc*.

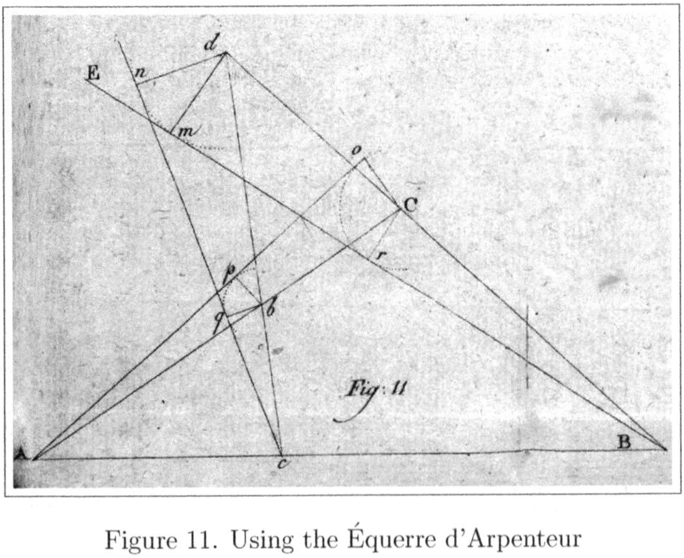

Figure 11. Using the Équerre d'Arpenteur
to prolong a line beyond an obstacle, Solution

[70] This is an application of what we have said (N.◦ 17).

This Solution does not absolutely require that the lines *dm*, *Cr*, *dn*, *bq*, *Co*, and *bp* are perpendicular two by two on *Bd*, *cd*, and *Cb*. Obviously is suffices, just the first two, the second two, and the last two are parallel to each other. Thus, the Solution may be provided by any instrument that provides a means of transporting an angle from one point to another.

53. We may also solve the same problem in a different way.

DE is the line that that we wish to prolong and which we see the points *D* and *E*. With the Équerre d'Arpenteur, I am looking for with the Équerre the point *B*, where I can see the points *D* and *E* in the two branches of the instrument. To do this, I take an arbitrary direction *BE* on which I walk holding one branch directed at *E* until I see the other point *D* in the other branch. I am looking for a second point *C* like that of *B*, that is to say, such that the angle *ECD* is right. Having drawn *CB*, I form, by any means whatsoever (this means will be very simple, if the instrument is composed of two mobile allidades around a center) on *BE* the angle *EBA* equal to the angle[54] *EBa*, [71] and on *CE* the angle *ECA* equal to the angle *ECB*, and the two lines *BA* and *CA* intersect at the point *A* on *ED*. Fig. 4.b

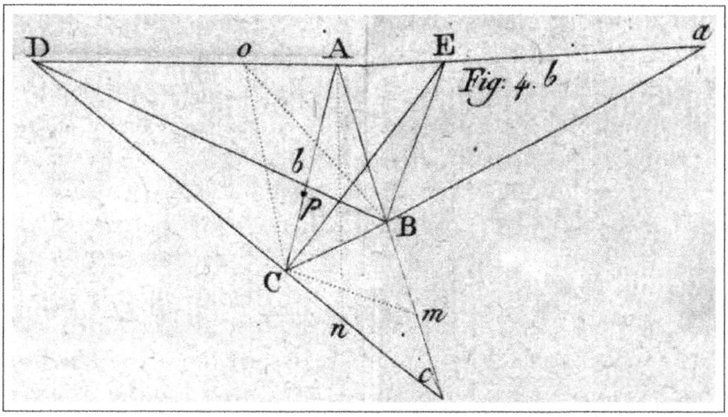

Figure 4.b. Using the Équerre d'Arpenteur
to prolong a line beyond an obstacle, First Alternate Solution

[54]The point *a* in Figure 4.b has not been defined; however, $EBa = EBC$.

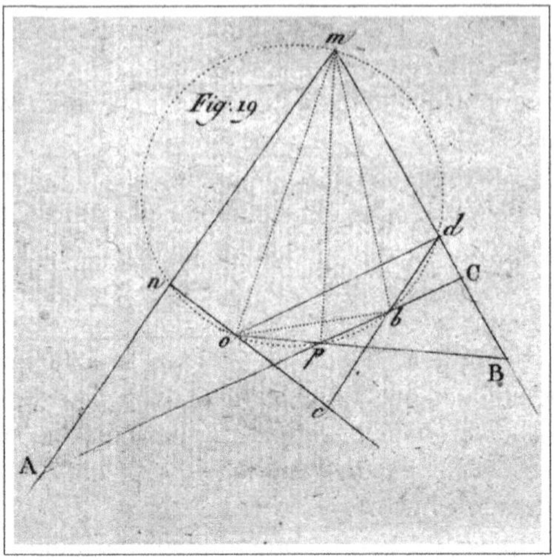

Figure 19. Using the Équerre d'Arpenteur
to prolong a line beyond an obstacle, Second Alternate Solution

This Solution clearly follows from what have demonstrated above (N.◦ 8).

Fig. 19 Here is a third Solution to the same problem, given equally by the Équerre and the fausse Équerre.[55] *Bc* is the line that we wish to prolong and which visibly presents the two points *B* and *c*. I place at will two Stakes at *d* and at *o*, which determine the alignments *dB*, *dc*, *oB*, and *oc*. I walk the instrument along *dc* until I find the point *b*, the vertex of the angle *dbo*, equal to one of the two given angles on the same instrument. I assume that this is the obtuse angle (if one uses the Équerre, this distinction is superfluous, because the Équerre gives two equal angles), also on *oB* I determine the point *p*, vertex of the angle *dpo*, also equal to the obtuse angle of the instrument. Then from the other side of the line *do*, on the lines *Bd* and *co* prolonged, I determine the points *m* and *n*, the vertices of the angles *dmo* and *dno*, both equal to the acute angles of the instrument. Finally, I place a Stake at *A* the point of intersection of the alignments *bp* and *mn*. Additionally, this Stake will be on the line *Bc*, [72] because, by the construction, the points *d*, *b*, *p*, *o*, *n*, and *m* are on the circumference of a circle, and give an inscribed hexagon, whose opposite sides intersect at the points *B*, *c*, and *A*. Therefore (N.◦ 17) these three points are on a straight line.[56]

---

[55]See p. 3, footnote 10.

[56]This is Pascal's Theorem after Blaise Pascal (1623–1662).

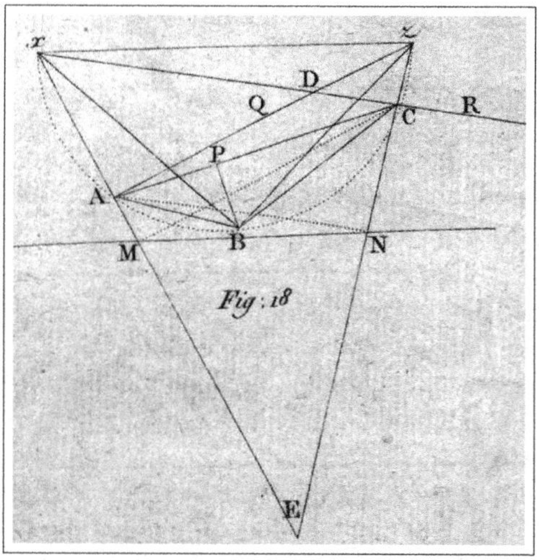

Figure 8. Using the Équerre d'Arpenteur to draw a parallel to an inaccessible line and to drop a perpendicular to an inaccessible line, Solutions

54. Solving the sixth problem with the Équerre.

xz is the inaccessible line and M is the given point. I determine, with the instrument, the points A and C, in such a manner that the angles xAz and xCz are right, and that the line Ax passes through the point M, which is always possible. I mark the point E of intersection of the two lines Ax and Cz. I measure AE, CE and EM. I then make Fig. 18

$$EN = \frac{EM \cdot AE}{CE},$$

and the line MN is the parallel sought. This Solution is indicated by Mascheroni (Book 2, problem III, Seventh Solution).

By the same construction, having marked the point D, where the two lines Az and Cx cross. If the point Q through which the parallel should pass is on AD, which we can always make, then in constructing on DC a line DR, which is the fourth proportional,[57] to DQ, AD, and CD, we will again have, the parallel sought through the points Q and R.

[73] It should be noted that for the Solution, it is not necessary that the xAz and xCz are right. It is enough that are equal to each other, and consequently the fausse Équerre may be used to determine them (I understand fausse Équerre, a similar instrument to the Équerre, but of which branches do not form a right angle). Given this, we can always make a circumference pass through the four points x, z, A, and C, and it

[57]The fourth proportional of x, y, and z is the value of t such that $x : y :: z : t$ or $\frac{x}{y} = \frac{z}{t}$.

is clear that the angle ACE, formed by AC and the prolongation of Cz, is equal to the angle Axz, because the measure of both them is half of the arc ACz. For the same reason, the angle CAE is equal to the angle Czx, was also clearly have the angle ACD equal to the angle Azx, and the angle DAC equal to the angle Cxz. However, from the construction of EN, we have the angle EMN equal to the angle ECA, and consequently equal to the angle Exz. Therefore MN is parallel to xz. It is the same reasoning that establishes the parallelism of QR.

However, without measuring any line and without any calculation, the Équerre or fausse Équerre give the parallel sought, by proceeding as follows. I first determine with the instrument the points A and C, as above, that is to say, in such a manner that [74] the angles xAz and xCz are equal. I place two Stakes at the points A and C – on the prolongations of Ax and Cz – I determine with the same instrument the points M and N such that the angles AMC and ANC are equal to one another and to the angles xAz and xCz. Additionally, the line MN is parallel to xz. Furthermore, it is accessible. Thus it only remains draw through the given point the parallel MN, which is a known practice with the instrument, without measuring a line or calculation.

Indeed, the four points A, C, M, and N are, by the construction, inscribable in a circle. Thus, the angle MNE is equal to the angle MAC; however, we have seen above that the angle MAC is equal to the angle Czx. Therefore the latter is equal to the angle MNE, and MN is parallel to xz.

55. Solving the thirteenth problem with the Équerre.

Fig. 18 Let xz be the line that we see the points x and z, and E the point from which we want to draw the perpendicular. I determine with the instrument, the point A on Ex, and the point C on Ez, in such a manner that the angles EAz and ECx are right. I mark the point of intersection D on the two alignments Az and Cx, and the direction ED is that of the perpendicular sought (N.◦ 6).

[75] At the point E, if I raise a perpendicular on ED, it will be parallel to xz. This is yet another way to draw a parallel to an inaccessible line, or to solve the sixth problem.

56. Solving the fourteenth problem with the Équerre.

Fig. 15 Az is the line that we wish to measure, which has the inaccessible extremity at z. The point A, with the instrument, I raise a perpendicular. I measure the arbitrary length on line AH, however, comparable to the presumed length of Az. Additionally, I place the instrument at H, one branch directed towards z, and I mark the intersection K of the alignment provided by the other branch with Az. I measure AK, and we have, as we know,

$$Az = \frac{\overline{AH}^2}{AK}.$$

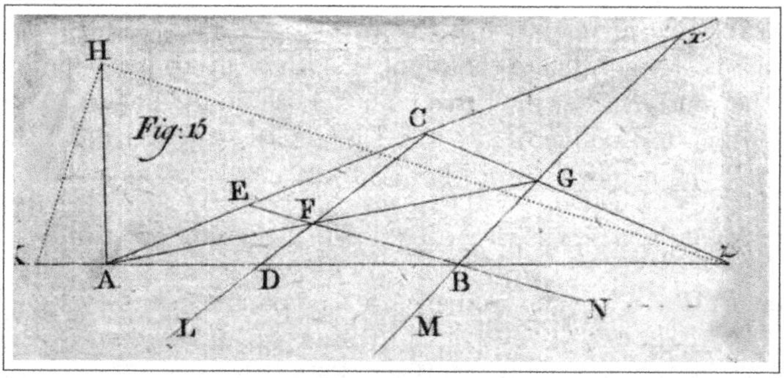

Figure 15. Using the Équerre Arpenteur to measure a line where one of the extremities is inaccessible, Solution

If we make the angles HAz and KHz equal to each other, and we measure AH, AK, and HK, we will have, because of the similar triangles AHz and KHz,

$$Az = \frac{AK \cdot \overline{HK}^2}{\overline{HK}^2 - \overline{AH}^2}.$$

57. Solving the fifteenth problem with the Équerre.

[76] xz is the line that we wish to measure. I choose three points A, B, and C, with the instrument, such that the angles xAz, xBz, and xCz are right, and I have the triangle ABC. with the same instrument I drop to the base AC, a perpendicular BP, from the vertex B, and after having measuring the lines AB, BC, and BP, I have Fig. 18

$$xz = \frac{AB \cdot BC}{BP}.$$

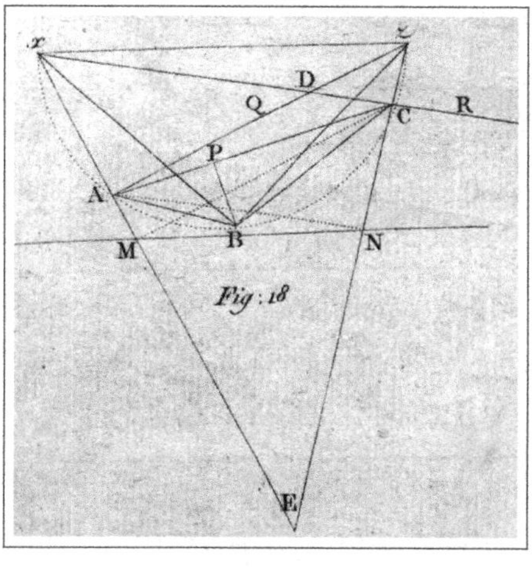

Figure 18. Using the Équerre d'Arpenteur to measure an inaccessible line, Solution

This very elegant Solution is due to Mascheroni (Book I, problem III, Solution 13). I restore here the proof that the Author has suppressed.

By the construction, the three points A, B, and C belong to a circumference which has diameter xz. Let the angles $Czx = m$, and the angle $Azx = n$. The angle AzC will be $= m - n$ in the triangle ACz, we will have

$$AC : \sin(m-n) :: Cz : \sin(CAz);$$

however, $CAz = Cxz$, and we have, in the triangle Cxz,

$$Cz : \sin(Cxz) :: xz : 1;$$

and in taking the radius in the tables equal to unity. Therefore,

$$xz = \frac{AC}{\sin(m-n)}.$$

Now, if we note that the angle ABx that is equal to the angle $Azx = n$; that the angle $xBz = q$ is a right angle, and consequently the angle

$$ABC = n + q + q - m = 2q - (m-n).$$

[77] We will see that $\sin(m-n) = \sin(ABC)$, and thus

$$xz = \frac{AC}{\sin(ABC)}.$$

However, in the right triangle BPC we have

$$\sin(BCP) = \frac{BP}{BC}.$$

In the triangle ABC we have

$$AC : \sin(ABC) :: AB : \sin(BCP) \text{ or } \frac{BP}{BC}.$$

Therefore,

$$\sin(ABC) = \frac{AC \cdot BP}{AB \cdot BC}.$$

Putting this value in the equation

$$xz = \frac{AC}{\sin(ABC)},$$

we have

$$xz = \frac{AB \cdot BC}{BP}.$$

We must also remark here that it is not necessary that the angles xAz, xBz, xCz, and BPC are right. It is sufficient that they are equal to each other. This is an observation that appears to have eluded Mascheroni. Indeed, let the angle[58]

$$xAz = xBz = xCz = BPC = o,$$

noting that after the construction, we can always pass a circumference through the five points x, A, B, C, and z, and reasoning absolutely as above, [78] we find

$$xz = \frac{AC \cdot \sin(o)}{\sin(m-n)}.$$

However, the angle $m-n$ or AzC, has as its measure half of the arc ABC, and the angle ABC has as its measure half of the remaining circumference. These two arcs are the supplements of each other, and consequently $\sin(ABC) = \sin(m-n)$. Furthermore, we have, in the triangle BPC,

$$\sin(BCP) = \frac{BP \cdot \sin(o)}{BC}.$$

Therefore for the triangle ABC, we have

$$\sin(ABC) = \frac{AC \cdot BP \cdot \sin(o)}{AB \cdot BC}.$$

Consequently,

$$xz = \frac{AB \cdot BC}{BP},$$

as above. The Solution may therefore be provided by the fausse Équerre.

[58] In [43, pp. 76–78], the use of o is meant to be an arbitrary angle, not zero.

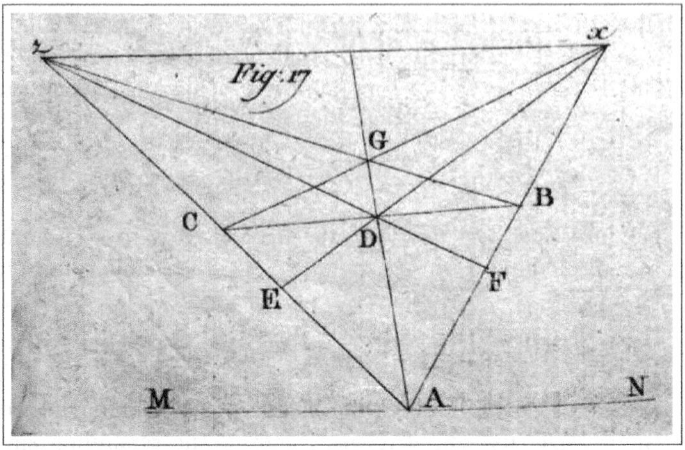

Figure 17. Using the Fausee Équerre
to measure an inaccessible line, Solution

58. It is not without a purpose that we have so often observed that the Équerre d'Arpenteur could replace the fausse Équerre, for you can find or procure it anywhere without expense. Two saw cuts that cross on the head of a picket make up the instrument, which may be often helpful, as we have seen. We can even help us to raise perpendiculars. For example, is it a matter of [79] dropping, from a point A that can be inaccessible, a perpendicular to a given line?

Fig. 17 xz is the given line. I mark with the instrument the two points z and x, from which I see the point A at equal angles zxA and xzA. Through the point C, taken at will on Az, I draw the parallel CB to zx. To do this we need take only one position with the instrument. I mark the point of intersection G of Cx and Bz, and the line AG will be the perpendicular sought. If we had wished instead, we could have taken the midpoint of the interval zx, it had been the foot of the perpendicular. However, this is what gives the construction without measuring anything (N.◦ 33).

Do we wish to raise a perpendicular on MN at the point A? I make the equal angles xAN and zAN. I then position the instrument twice at B and at x, to get both parallels BC and xz to MN. I mark the point of intersection G of the lines Bz and Cx, and AG is the perpendicular sought.

In Metz from the Printer d'Antoine l'aîné.

Appendix A

Letter from S... To F..., Professor of Mathematics

[1][1] I Thank you for the communication that you were good enough to send me of your remarks on the *Traité analytique des courbes et surfaces du second Ordre*,[2] published in Paris in 1802. A work that you have adopted, after the example of a large number of good Teachers, as the text for your lessons to young People that you are preparing to to enter the Ecole Polytechnique. I agree to examine with you the articles that you denounce as susceptible, otherwise censorship. To shorten the discussion, I will not follow the order point in which you have extracted them, It will be shorter to gather together for which the difficulties are derived from the same cause. I start with those difficulties properly belong to the principles of analytic Geometry.[2]

1. In analysis we make use of principles the truth of which we do not truly sense. These are those that concern the geometric effect of combinations of analytic formulas: we add, subtract, multiply equations, we eliminate; and we say tersely that the results represent this or that thing. Here you say, where the Student remains amazed, and where the Teacher stammers.... That is to say, the former understands nothing and the latter lacks the words to explain what he means.

These principles, which are not evident, are not as numerous as you seem to insinuate. I dare to assure you that, in the *Elements*,[3] there is but one, that is the one that pertains to what you call the *Geometric Effect* of the elimination. I also wish that it was expounded in the preliminaries with some scope, which does not hide the fact that it rests on very

[1]In [43, p. 1], the page numbers start from 1 at the start of this letter.

[2]This is reference [4] in the current bibliography.

[3]In [43, p. 1], this title was not italicized. This is reference [17] in the current bibliography.

subtle considerations, that we even advised, if necessary in this regard, d'Alembert[4] recipe for obtaining faith in mathematics. The principle, declared in this way, vividly seizes the minds of young people. It would be returned to with curiosity, dissected.... [3] Additionally, if it is true that it takes time and effort to feel the evidence for it, I believe that the trial would not be of very long duration.

However, we must prove that the difficulties of the principles in the *Elements*, in general reduce to the principle of elimination.[5]

In the circle, the equations of two cords through the extremities of the diameter are $y = a\,(x + R)$ and $y = a'\,(x - R)$ (N.◦ 49, page 76).

> Suppose (says the Author) that these two straight lines must intersect each other on the circumference of the circle, it must be that their equations must hold at the same time and with that of this circumference. Now, in multiplying them, they give $y^2 = aa'\left(x^2 - R^2\right)$. Additionally, in order that this result is consistent with $y^2 = R^2 - x^2$, which is that of the circumference, we must have $aa' + 1 = 0$.

However, you ask, what does this second degree equation, the result of the multiplication of two linear equations? Why do we multiply instead of adding or subtracting, or form other combinations? Does analysis have its operations of *construction* [4] as synthesis has its lines of *construction*?

I reply, first of all, that in admitting the combination of these two linear equations, we suppose that the variables x and y have common values in both of them, and consequently the lines they represent intersect. Where as we do not assume a relationship between a and a', the equation resulting in the combinations represents either a line passing through the intersection of the first two, or else the system of two lines, or etc.. It is also the same thing, if, in establishing a relation between a and a', we did not make one of these parameters disappear from the combination by means of this relation. However, it by means of the relation, we eliminate the parameters a and a' from the equation, the result is the equation of the line determined by the intersections of the two straight lines in all positions that they can take following the assumed relationship. If, for example, we subtract the two equations, and we make

$$\frac{a}{a'} = b,$$

b being a constant, the result

$$x = -R\left(\frac{b+1}{b-1}\right)$$

[4]Jean le Rond d'Alembert (1717–1783).

[5]In other words, using an addition elimination method to solve two systems of equations.

is the equation of the continual intersection [5] of two lines taking all positions consistent with the relationship

$$\frac{a}{a'} = b;$$

however, this equation of intersection is a straight line parallel to the y-axis. Additionally, because we wish to compare here the location of the intersections with the circumference of the circle, this combination is useless. By multiplying the two equations, and eliminating these two parameters by the relation $aa' = c$, we have an equation of the second order $y^2 = c\left(x^2 - R^2\right)$, which being a general conic section, becomes a circle identical with the one proposed by making $c = -1$. If we first assume between a and a' the relation that $aa' + 1 = 0$, which expresses that the two cords are always perpendicular to each other. The equations being $y = a\,(x + R)$ and

$$y = -\frac{1}{a}\,(x - R)\,,$$

and eliminating them, which is equivalent to multiplying them, we would obtain the equation of the circle $y^2 = R^2 - x^2$. Therefore, we see that there is here but a singular effect from the elimination of the parameters a and a' between the equations of the cords and the [6] relation established between them, and that multiplication was only the means to produce this elimination. Here is one difficulty already reduced to the principle of elimination. The articles (58, page 92, etc.) on additional questions of cords to the ellipse and the hyperbola, give rise to the same remarks.

The equation of the tangent to the circle was found to be (N.◦ 52, page 82) $yy'' + xx'' = R^2$, where[6] (x'', y'') is the point of tangency.

> We have supposed (says the Author) up till now the point of tangency to be given, let us regard it now as being unknown, and we propose to determine in such away that he tangent passes through a given point outside the circle. Let x' and y' be the coordinates of this point, because it must be on the tangent, it satisfies the equation of this line, which requires that we have $y'y'' + x'x'' = R^2$. We also have $y''^2 + x''^2 = R^2 \ldots.$ Subtracting the two equations from each other, they give $y''^2 - y'y'' + x''^2 - x'x'' = 0 \ldots.$ The coordinates x'', y'' belong to a circumference... (which)... will cut the given circumference in two points [7] which will be the points of tangency that we seek.

Here, you say, it is no longer multiplication that carries out the magical effect, but subtraction. As for me, I once again see it clearly as elimination,

[6] In [43, p. 6–11], the notation used for a coordinate points was of the form x, y; however, the modern convention (x, y) will be utilized throughout.

because the subtraction has removed the R^2 from the two equations and does nothing else, however, what does the second degree of the equation signify? Its locus is a curve on which are situated all points of contact of tangents that maybe drawn through the point (x', y') in the infinite sequence of circumferences concentric with the opposed circumference, or, to speak more precisely, the equation is a curve of the location of the continual intersections of the line $y'y'' + x'x'' = R^2$ with the circle $y''^2 + x''^2 = R^2$, for whatever radius R. However, what is this line $y'y'' + x'x'' = R^2$, when x'' and y'' are regarded as variables? First of all it is a straight line. Its equation is only of the first degree. Furthermore, this equation is verified by the particular values y'' and x'', the coordinates of the points of contact of a tangent drawn through the point (x', y'). This line, therefore, passes through these points; it is therefore a cord that joins it. Thus, by constraining it, we have, by its [8] intersection with the circle, the points of contact sought. From this it follows that, in order to solve the problem that we proposed to ourselves, the subtraction and its results are not even helpful. Indeed, if, through the center and the point (x', y'), we draw a straight line, its equation will be

$$y = \frac{y'x}{x'}.$$

Then the equation of the cord gives

$$y'' = -\frac{x'}{y'}x'' + \frac{R^2}{y'}.$$

Therefore, these two lines are perpendicular to each other. Furthermore, when $x'' = 0$, we have

$$y'' = \frac{R^2}{y'}.$$

Therefore, if we construct the quantity

$$y'' = \frac{R^2}{y'}$$

on the y-axis, and if we drop from the vertex y'' a perpendicular on the line that joins the center and the point (x', y'), this perpendicular cuts the circle at the two points of contact. Alternately, if we place the line that joins the center and the point (x', y') on the x-axis, which does not change the form of the equations of the circle and of the cord, but only gives $y' = 0$, the equation of the cord becomes $x'x'' = R^2$; from which

$$x'' = \frac{R^2}{x'},$$

the value that indicates the same construction [9] as the equation of the second degree obtained by the elimination of R^2.

However, you say, the equation of the tangent of the ellipse, passing through the point (x', y'), is

$$A^2y'y'' + B^2x'x'' = A^2B^2;$$

the equation of the curve being

$$A^2y''^2 + B^2x''^2 = A^2B^2$$

(N.◦ 64, page 100). If we subtract these two equations, we have the equation as the result

$$A^2y''^2 - A^2y'y'' + B^2x''^2 - B^2x'x'' = 0,$$

in which we have not eliminated A and B, nevertheless we also say that this result is of a curve which through its intersections will give the points of contact from the tangent drawn through the point (x', y'). It is true that, by the subtraction, we only eliminate A^2B^2; also that the result is the equation of the curve of the location of the continual sequence of the points of contact of the tangents drawn from the point (x', y') to the infinite sequence of ellipses, not only concentric, but similar to the given ellipse, that is to say those for which the ratio of the axes is the same. For let

$$\frac{A}{B} = D$$

be a constant quantity; after the elimination of A^2B^2, D persists in the result, while A and B have disappeared.

Here, as with the circle, the equation

$$A^2y'y'' + B^2x'x'' = A^2B^2$$

is clearly [10] that of a line which joins the two points of contact for the tangent drawn through the point (x', y'). Additionally, this is still true when the ellipse is related to its conjugate diameters. That is to say that the equation

$$A'^2y'y'' + B'^2x'x'' = A'^2B'^2$$

(N.◦ 124, page 233) is still that of the cord drawn through the points of contact of the tangent drawn through the point (x', y'). If we make a diameter pass through the point (x', y'), we will have $y' = 0$, and the previous equation therefore being

$$B'^2x'x'' = A'^2B'^2,$$

we will have

$$x'' = \frac{A'^2}{x'}$$

a quantity easy to construction by the circumference of a circle. Therefore, drawing through the extremity of x'' an ordinary parallel to the conjugate diameter, we will have two points of contact on the ellipse, and from there by a method that is simple and elegant a method of drawing tangents to the ellipse through an exterior point. Analogous observations hold for the hyperbola.

You will permit me, at this point, to remark in passing, that the equation of the normal of the ellipse being

$$B^2yx'' - A^2xy'' = -\left(A^2 - B^2\right)x''y''$$

(N.◦ 61, page 97), If we make this line pass through any point (x', y'), we will have

$$B^2y'x'' - A^2x'y'' = -\left(A^2 - B^2\right)x''y''.$$

[11] Additionally, if we suppose x'' and y'' to be variables in this, we will have the equation of a line passing through the points of the ellipse where any line drawn through the point (x', y') would be normal. However, this equation is a hyperbola that is easy to construct, if the given point is on the first axis,[7] we have $y' = 0$; from which $y'' = 0$ and

$$x'' = \frac{A^2x'}{A^2 - B^2}.$$

The equation $y'' = 0$ indicates that the major axis itself passes through the two normal points. The latter, which is a perpendicular to the major axis, indicates two other normal points, which will occur when the value of x'' is not as large as A, which happens when x' is smaller than

$$\frac{A^2 - B^2}{A}.$$

I come to articles where it is specifically a matter of the question of elimination, they are, among others, those located by the Author of the equation of the plane (N.◦ 33, page 50); the equation of the surface of the right circular cone (N.◦ 43, page 67); the intersection of two planes (N.◦ 36, page 59), etc.

You had wished that we had given the equations of the plane and the conic surface, following the methods of the ancients which, as you say, are much easier to explain. I [12] am not of your opinion, and I think we would do well to put an Essay on the method, as fecund as it is elegant, in deducing the equations of surfaces from their generating lines in the hands of Novices. However, I lament as you do the terseness that has been used in the exposition of this method. In my opinion, it would be here that we should solemnly proclaim the full extent of the principle of elimination,

[7]That is, the x-axis.

its influence on all parts of analysis, and to accord it at least a few lines of exposition.

Here is how I have sometimes organized the examination of the principles relating to elimination. My method is no doubt far from being perfect. However, it has the advantage of having distinct images in their details.

If we have the two equations $\mathrm{f}(x, y) = 0$ and $f(x, y) = 0$, (I use these well-known symbols for brevity) which are two plane curves – assuming that they both hold – we show that they have common points. If we eliminate the variable x from them, we have $\mathrm{F}(y) = 0$ as results, whose roots are the ordinates of the common points. If we eliminate y from the [13] same equations, we have $F(x) = 0$, whose roots are the abscissas of the common points. Thus, we have everything it takes to determine their position. This is without difficulty.

If we have the two equations $\mathrm{f}(x, y, z) = 0$ and $f(x, y, z) = 0$, which are two curved surfaces – assuming that they both hold – we shove that they intersect each other. By eliminating the variable x from them we have $\mathrm{F}(y, z) = 0$, the equation of a plane curve in the yz-plane, which is the projection on this plane, of the intersection of the two surfaces. We can't add anything to the illuminating exposition of this principle by the celebrated Author of the *Géométric descriptive.* (Art. 49, page 60.)

If we have the equations $\mathrm{f}(x, y, a) = 0$ and $f(x, y, a) = 0$, a being constant, which again are two plane curves, we find that they intersect each other by making them hold together. By eliminating the constant a from them, we will have $\mathrm{F}(x, y) = 0$. An equation whose locus is the continual sequence of the intersections, two by two, of the curves we obtain in varying the parameter a indefinitely in the given equations. Indeed, a maybe regarded as one of the particular values [14] of z in the equations $\mathrm{f}(x, y, z) = 0$ and $f(x, y, z) = 0$, which are the two curved surfaces, and their two proposed equations will be the projection on the xy-plane of the sections made in the two curved surfaces by a plane parallel to the xy-plane and elevated above it by a height equal to a, or, if we wish, those of the sections described themselves by the consideration of the intersecting plane. Furthermore, it is clear that the points where these curves intersect belong to the intersection of the two curved surfaces. If we imagine an indefinite sequence of intersecting planes, parallel to the first, they will each give two sections, which will be represented on them by the two given equations, by putting into each of them the distance for each of the xy-planes, and in each plane these curves intersect at the intersection of two surfaces. However, by eliminating a in the two equations, it is as if we eliminated z in the the equation of the two surfaces. Now, by the preceding principle, the result of the elimination of z, namely $\mathrm{F}(x, y) = 0$ is the equation of the projection on the xy-plane of the intersection of the two surfaces. It is therefore also the equation of the projection in the same plane of all the [15] points where the two imagined sections intersect, two by two, in their respective planes. Similarly, if we pull back each of these planes themselves to the xy-plane, in the direction perpendicular to it, all of these points of intersection will arrange themselves in the same plane, on the perimeter of the curve, of which $\mathrm{F}(x, y) = 0$ is the equation. That is to say, that we may consider this curve as being formed by sequence of these points of intersection, or, which is the same thing, as being formed by the result of the sequence of intersections of the curves that we obtain on xy-plane, by varying a in the two given equations.

If we have the equations $\mathrm{f}(x, z, a) = 0$ and $f(y, z, b) = 0$, which are the equations of a line of double curvature,[8] and at the same time the equation $\mathrm{F}(a, b) = 0$, and we eliminate a and b by this third equation, we will have $F(x, y, z) = 0$, the equation of a curved surface generated by the motion of a line determined by the relation rule $\mathrm{F}(a, b) = 0$. That is the principle expressed generally. I will consider it in particular cases where one finds its application in the *Elements*.

Let $y - y' = a\,(z - z')$ and $x - x' = a'\,(z - z')$ be the equations of a straight line that passing through [16] the point (x', y'), and such that the position is otherwise determined by the quantities a and a'. If the straight line changes position around the point (x', y'), following a certain law,[9] it must be the case that there is a certain relation between the quantities a and a' that I designate by $a = \mathrm{f}(a')$. For the moment, I suppose x is constant and through its extremity is a plane parallel to the yz-plane. The two equations being those of the projections of straight lines on the yz-plane and xz-plane, the coordinates (y'', z'') of the point of intersection satisfy them in one of their particular positions, being the plane we supposed, and we have[10]

$$y'' - y' = \mathrm{f}\,(a') \cdot (z'' - z')$$

and

$$x - x' = a'\,(z'' - z')\,,$$

equations in which everything is determined. We may thus take a particular value of a' in the second, namely:[11]

$$a' = \frac{x - x'}{z'' - z'},$$

and transporting it in the first equation, which will give

$$y'' - y' = \mathrm{f}\left(\frac{x - x'}{z'' - z'}\right) \cdot (z'' - z')\,.$$

[8]In other words, a non-developable surface. A non-developable surface is one that cannot be flattened without being dilated. See [57, p. 311–12].

[9]These are conic sections.

[10]In [43, p. 16], the "and" was missing between the following two equations.

[11]The value of a' was given as $z'' - z' \frac{x - x'}{z' - z''}$. [*Ibid.*]

By giving another value to a', and consequently another position for the straight line, we will find by the same method, by designating by (y''', z''') the coordinates of its intersection in the plane, the equation

$$y''' - y' = \mathrm{f}\left(\frac{x - x'}{z''' - z'}\right) \cdot (z''' - z'),$$

[17] which is the same as the previous in the accents[12] attached to the variables y and z. Therefore, in suppressing these accents, the equation

$$y - y' = \mathrm{f}\left(\frac{x - x'}{z - z'}\right) \cdot (z - z')$$

is that of the given curve, in the plane we supposed, by the continual intersections of the straight line indefinitely changing position, where the equation of the section made in the plane, in the surface generated by the movement of the line. Therefore also by supposing x as variable, the same equation must be that of the surface itself. However, it is clear that the equation is the one we would obtain by eliminating a among the three given equations.

It was only after an extended discussion of these specific principles that we may, as a kind of induction, conclude that in general

> The elimination of certain quantities among the equations, true together, makes the other quantities independent of the other particular values that the eliminated quantities may take, and brings us to results that collectively include all those that we had separately for each possible value of the eliminated quantities.

[18] It is more or less in these terms that we find the principle in various works. (See Lacroix[13], volume 1, page 395, note.)[14]

2. I turn to the articles where you have noted difficulties of another kind. As you say they are not as serious as the first and certainly they would never stop a good mind in its path. For example...

It is a matter (N.◦ 67, page 104) of proving that every diameter in the ellipse, divides the cords parallel to the tangent drawn through its extremity into two equal parts, and for this we make use of a transformation of the equation of the curve that you compare to the most complicated synthetic constructions. However, it is not necessary to load the memory of the Student for all of this calculation. The equation of a cord passing through through the point (x''', y''') of the diameter being $y - y''' = a\,(x - x''')$. If

[12] That is, the primes.

[13] Sylvestre François Lacroix (1765–1843).

[14] *Traité du calcul différentiel et du calcul intégral.* This is reference [26] in the current bibliography.

this straight line meets the ellipse at the points $('x,' y)$ and $(''x,'' y)$, we will have at the same time

$$a = \frac{'y -'' y}{'x -'' x}$$

and

$$A^2 \left('y^2 -'' y^2\right) + B^2 \left('x^2 -'' x^2\right) = 0,$$

from which we immediately conclude

$$\frac{'y -'' y}{'x -'' x} = -\frac{B^2 ('x +'' x)}{A^2 ('y -'' y)} = a.$$

However, when the [19] cord is divided into two equal parts by the diameter we clearly have $'y +'' y = 2y'''$ and $'x +'' x = 2x'''$. Therefore,

$$a = -\frac{B^2 x'''}{A^2 y'''},$$

as we see in the location cited. We may apply the same observation to the transformed equation indicated (N.◦ 78, page 125).

You find the method given (N.◦ 80, p. 127) for obtaining the surface of the parabola too difficult for Novices. I regard it only as unwarranted, because it requires developments that the young people would see more opportunely in studying the *Théorie des fonctions analytiques*,[15] and because we do not apply it to the quadrature of the other curves of the second order. I prefer, to this end, the method which we get from the comparison of elementary rectangles $y'(x'-x)$ and $x'(y'-y)$, their ratio is

$$\frac{y'(x'-x)}{x'(y'-y)} = \frac{y'(y'+y)}{2px'} = \frac{y'+y}{y'}.$$

Now, we see that the closer y' comes to being equal to y the more this ratio approaches $\frac{2}{1}$, which consequently the limit of the first quantity. Therefore, the ratio $\frac{2}{1}$ is also that of the area of AMP to [20] the area of AMR, the complement of the first in the rectangle xy (fig. 49); that is to say that the parabolic area AMP is two-thirds that of the rectangle xy. This path is also more analogous to the method by which the Author has found (N.◦ 69) the ratio of the elliptical area to that of the circle described on the principle axis.

You found obscurity in the manner by which we assign (N.◦ 94, 2.nd *discussion*, etc.) the conditions that always make the variable y imaginary in equation (1). A word will make that disappear. In order that y always imaginary, it must be the case that is necessary that, regardless of x, the polynomial under the radical is negative. Therefore $B^2 - 4AC$ must

[15] This is reference [27] in the current bibliography.

already be negative, because otherwise we could give to x a value larger that the polynomial to became positive. This being, it is necessary that the polynomial, divided by $B^2 - 4AC$ gives a positive result no matter what x is (x always being real). However, the latter condition is fulfilled, if the polynomial equal to zero gives x as imaginary, because we know that it is a property of equations that have all their roots imaginary to always gives positive results, no matter what value we substitute for the unknown.

I do not see any better than what use equation (2) (page 152) might be coming to the transformation that should confirm that, *a posteriori*, that what we have just established. Equation (1) alone brings us there easily, because after the radical disappears it is sufficient to form a perfect square using the two terms x.

It is a matter (N.◦ 95, page 153 and following) of demonstrating that any second degree equation with two variables and in rectangular coordinates represent a conic section or represent nothing. Under the pretext greater simplicity in calculation, the Author gives to the cone a particular position of having one of its generators perpendicular to the cutting plane. By means of this, it would seem to you that we only could have concluded, at the end of the calculation, any equation of the second degree represents any section made in a cone by a plane cutting it perpendicularly to one of its generators or representing nothing at all. However, you say, may we also thereby concluded that it is impossible to analytically represent a section made in a cone by a plane cutting obliquely to all generators, which would be absurd. Be careful, he [22] has slipped a sophism in your way of reasoning, because there are six indeterminates in the general equation of the conic sections: namely, M, a, b, x', y' and z'. Additionally, in a given complete equation of the second degree, there are indeed six coefficients, however, one of them is arbitrary, because any equation may be multiplied by an arbitrary factor, and, in the comparison of this with the first observation, we may always have dispose of the arbitrary factor to establish a particular relationship between the indeterminates of the equation of the conic section. This is what we have done here. We have to disposed of the factor to give the relation $M^2\left(1 + a^2 + b^2\right) = 1$. Therefore, the legitimate conclusion is that any second degree equation with which we wish to represent a section made in the cone by a plane perpendicular to a generator represent only them exclusively, yet this does not prevent that we be able to determine the arbitrary factor by introducing any other relation between the indeterminates M, a, b, x', y' and z'. However, in admitting other relation, that is to say other positions of the cone with respect to the cutting plane, can not the comparable second degree equation represent anything else but these [23] new sections? It is this that the Author does not examine, and that is because, in my opinion, his demonstration is not complete. To this end, we may give convenient generality to the demonstration by making $M^2(1+a^2+b^2)=G$, G being a quantity given by a particular comparison, but, variable from one to the other. Given this, so that the position of the cone not be imaginary, we must first of all by that G is a positive quantity, because it is equal to the product of two squares. Furthermore, it must also be that it is smaller than $1+a^2+b^2$, because M^2 is the square of a cosine, and consequently always smaller than unity. With these ideas, and in arranging the calculus roughly as the Author, we find that z' is only imaginary in the same circumstances that make the general second degree equation absurd.

When $B^2-4AC=0$ (N.◦ 98, page 162) the curve is a parabola, because then we may always draw two straight lines, one being parallel to x, and another being parallel to y, which each leave the curve entirely on the same side. However, it was said above (page 158) that the characteristic by which we recognize the parabola, is

> That the curve [24] only extends indefinitely in two directions, that of its axis and that of the lines to which it is perpendicular.

From this it follows, you say, that we may draw a straight line that leaves the entire curve on one side perpendicularly, but not parallel to the axis of the curve, and consequently that there is a contradiction between the two passages. The contradiction is only apparent, because in the first section, we suppose, *tacitly* that it is true that A and C are not zero as it is easy convince one self, whereas in the latter, where it is a matter of a parabola with respect to its axis, we suppose $A=0$ or $C=0$ at the same time as B. That is to say, that there is such a system of coordinate axes that includes the entire parabola in one of its angles, and another system that permits it to extend in two adjacent angles.

You find the application of the transformation of the coordinates to the discussion of the general equation of the second degree much too diffuse. It occupies forty pages (from N.◦ 100 to N.◦ 119).

It is true that we could have restrained it greatly. Here is the order in which I have sometimes [25] advised that we study this part of the work. First of all to change the position of coordinated axes, in assuming any new whatever inclination of the new axes between one another and with the originals, making use of this transformation to make the rectangle coordinates to disappear (N.◦ 118), to remark that the condition of equation obtained for it always gives real values for the new angles of the original axis, one of them remaining arbitrary, and that it is *singularly* satisfied by a system of two rectangular axes.

Transporting the new axes parallel to themselves by introducing indeterminates a' and b'. To dispose of the quantities a' and b', to make terms of variables of the first power to disappear, under the hypothesis that the new axes are rectangular (N.◦ 108 and N.◦109), which reduces to the consideration of the second order curves to their centers and their axes, and under the hypothesis of non-rectangular coordinate axes (N.◦ 115 and 116) to have the theory of conjugate diameters. Next, to dispose of the arbitrary a' and b' to make the known term and one of the first [26] powers of the variables disappear, an operation that places the origin on the curve and gives an equation that applies to curves of the second order that have a center as well as those who do not (N.◦ 112, N.◦ 119). Finally, to remark we may dispose of the the angles of the new axes with those of the the old ones, to make the squares of variables disappear, and by means of this to brought to the consideration of the hyperbola with respect to its asymptotes (N.◦ 110, N.◦ 117), and to complete it with examples and discussions (N.◦ 113). According to this distribution, which eliminates repetition and permits illuminating connections, we may say everything in a very small space.

The transformation that is used (N.◦ 121) seems to you unnatural and very complicated to you. You substitute others simpler ones and I can only applaud this.

We affirm (N.◦ 122) that the angle of the supplementary cords that intersect at the extremity of the minor axis of the ellipse is the greatest. You ask how we can demonstrate it without making use of the differential calculus. Here it is:

The trigonometric tangent of the angle of two supplementary cords is generally [27]

$$\frac{a'-a}{1+aa'},$$

and we have the relation

$$aa' = -\frac{B^2}{A^2},$$

a constant quantity. From this it follows, first of all that the first expresses increases as $a'-a$, or rather as $a'+a$, by precision consideration of signs, because by virtue of the relation, the two quantities are always have different signs. Let $m = a'+a$ and $n = a'-a$, we have

$$a' = \frac{m+n}{2}, \quad a = \frac{m-n}{2}, \quad \text{and} \quad aa' = \frac{B^2}{A^2} = \frac{m^2-n^2}{4},$$

from which

$$m^2 = \frac{4B^2}{A^2} + n^2,$$

and consequently m, or $a'+a$, is the smallest possible when $n = 0$, or $a' = a$. However, the angle under consideration being always obtuse, it will be all the more larger when the tangent is smaller, it will be greatest when we have $aa' = a^2$ or

$$a = \frac{B}{A},$$

that is to say when the two cords intersect at the top of the minor axis.

I close here a discussion that would be unnecessarily prolonged, because the other difficulties you raised are either not or are so minor that they do not deserve to be stopped at, and I conclude that there is very little missing in the work being examined to be shelter it from all criticism. [28] However, this is not enough, a mediocre work may be without significant defects. The impartiality requires that we recognize that it is at any rate very well done, and that it assembles together, in a very high-degree, the qualities that recommend a good elementary work: order, sharpness, precision, elegance, depth, new points-of-view, etc.

Appendix B

Table to the Practical Part

1. To prolong a line beyond an obstacle, N.os 19, 20, 21, 22, 52, 53.

2. To find a point aligned with the two invisible points of intersection of two pairs of lines given in direction, N.o 23.

3. From a given point, to draw a line to the invisible point of intersection of two lines given in direction, N.os 24, 25.

4. At a given point, to draw a line parallel to two other parallels given in direction, N.o 25.

5. To draw two parallel lines on the ground, N.o 26.

6. From a given point, to draw a parallel to an accessible line, N.o 27.

7. From a given point, to draw a parallel to an inaccessible line, N.os 28, 29, 30, 31, 54, 55.

8. To divide a line into two equal parts, N.o 33.

9. To divide a line into any number of equal parts, N.o 34.

10. To divide an accessible angle into two equal parts, N.º 35.

11. To divide an inaccessible angle into two equal parts, N.ºs 36, 37.

12. To erect a perpendicular on a line at a given point, N.ºs 39, 58.

13. From an inaccessible point, to drop a perpendicular to an accessible line, N.ºs 40, 58.

14. From a given point, to drop a perpendicular to an inaccessible line, N.ºs 41, 55.

15. To measure a line, one of whose extremities is inaccessible, N.ºs 43, 44, 56.

16. To measure an inaccessible line, N.ºs 46, 47, 57.

17. To measure the distance from a point to an inaccessible line, N.º 49.

Bibliography

[1] Adams, G. and Jones, W. (1803). *Geometrical and Graphical Essays, Containing a General Description of the Mathematical Instruments Used in Geometry, Civil and Military Surveying, Levelling, and Perspective: With Many New Practical Problems.* London: W. Glendinning.

[2] Aebischer, A. and Languereau, H. (2010). *Servois ou la géométrie á l'école de l'artillerie.* Besançon Cedex: Presses universitaires de Franche-Comté.

[3] Bernoulli, J. (1742). *Opera omnia.* Sumptivus Marci-Michaelis Bousquet & sociorum.

[4] Biot, J. B. (1802). *Traité analytique des courbes et des surfaces du second degré.* Paris: De l'imprimerie de Crapelet.

[5] Boyer, J. (1895). "Le mathmaticien franc-comtois François-Joseph Servois," *Mémoires de la Société d'émulation de Doubs* **9**, 305–13.

[6] Boyer, J. (1895). "Pièces Justicatives et Notes Diverses," (appendix to the foregoing, including documents from Servois' military dossier and notes by Abbé Filsjean). *Mémoires de la Société d'émulation de Doubs* **9**, 314–28.

[7] Boyer, C. B. (1949). *The History of The Calculus and Its Conceptual Development.* New York: Dover Publications.

[8] Boyer, C. B. (1989). *A History of Mathematics.* New York: John Wiley and Sons.

[9] Boyer, C. B. (2004). *History of Analytic Geometry.* New York: Dover Publications.

[10] Bradley, R. E. (2002). "The Origins of Linear Operator Theory in the Work of François-Joseph Servois," *Proceedings of Canadian Society for History and Philosophy of Mathematics* **14**, 1–21.

[11] Bradley, R. E. and Petrilli, S. J. (2010). "Servois' 1814 Essay on the Principles of the Differential Calculus, with an English Translation," *Loci: Convergence*, **7**.

[12] Bradley, R. E. and Petrilli, S. J. (2010). "Servois' 1814 Essay on a New Method of Exposition of the Principles of the Differential Calculus, with an English Translation," *Loci: Convergence*, **7**.

[13] Carnot, L. (1803) *Géométrie de position.* Paris: Durprat.

[14] Coxeter, H. S. M. and Greitzer, S. L. (1967). *Geometry Revisited.* Washington D. C.: The Mathematical Association of America.

[15] D'Antoni, A. (1777). *Institutions Physico-Mechaniques V1: A L'Usage Des Ecoles Royales D'Artillerie Et Du Genie De Turin.* Bauer & Treuttel.

[16] Del Latto, A. & Petrilli S. J. (2012). "Algebraic Formalism within the Works of Servois and its Influence on the Development of Linear Operator Theory," *Loci: Convergence*, **9**.

[17] Euclid. (2006). *The Elements.* New York: Barnes and Nobles. (Original work published in the fourth century B. C. E.)

[18] Français, J. F. (1813). "Chronologie. Solution directe des principaux problèmes du calendrier,"*Annales de mathématiques pures et appliquées* **4**, 273-276.

[19] Francoeur, L. (1801). *Traité de mécanique élémentaire à l'usage des élèves de l'école polytechnique.* Paris: H.L. Perronneau.

[20] Gillispie, C. (2004). *Science and Polity in France: The Revolutionary and Napoleonic Years.* Princeton, NJ: Princeton University Press.

[21] Grabiner, J. V. (1981). *The Origins of Cauchy's Rigorous Calculus.* New York: Dover Publications.

[22] Hamilton, W. R. (1853). *Lectures on Quaternions: Containing a Systematic Statement of a New Mathematical Method; of which the Principles Were Communicated in 1843 to the Royal Irish Academy; and which Has Since Formed the Subject of Successive Courses of Lectures, Delivered in 1848 and Subsequent Years, in the Halls of Trinity College, Dublin: with Numerous Illustrative Diagrams, and with Some Geometrical and Physical Applications, Volumes 2–4.* Hodges and Smith.

[23] Herman, C. (1992). *Knights and Kings in Early Modern France: Royal Orders of Knighthood, 1469–1715.* New York: AMS Press.

[24] Kästner, A. G. (1800). *Mathematische Anfangsgründe: Anfangsgründe der Arithmetik, Geometrie, ebenen und sphärischen Trigonometrie, und Perspectiv.* Vandenhoek.

[25] Krieger, L., Jantzen, S, and Neill, K. (1992). *World History: Perspectives on the Past.* Lexington: D.C. Heath and Company.

[26] . Lacroix, S. F. (1797). *Traité du calcul différentiel et du calcul intégral.* Paris: J.B.M. Duprat, 1797–1800. Tome premier.

[27] Lagrange, J. L. (1797). *Théorie des fonctions analytiques.* Paris: De L'imprimerie de la République.

[28] Lambert, J.H. (1759). *La perspective affranchie de l'embarras du plan géometral.* Zuric: Chez Heidegguer et Comp.

[29] Liechty, M. (2002). *The Revolutionary Calendar.* Retrieved June 4, 2008, from Research Subjects: Government and Politics - The Napoleon Series, Web site: URL.

[30] Mascheroni, L. (1793). *Problemi per gli agrimensori con varie soluzioni dell'ab.* Pavia: Presso Baldassare Camino.

[31] Mascheroni, L. (1803). *Problèmes de géométrie pratique pour les arpenteurs avec différentes solutions.* Paris: Chez Courcier.

[32] Mascheroni, L. (1838). *Problèmes de géométrie pratique pour les arpenteurs avec différentes solutions.* Paris: Bachelier.

[33] Miller, G. (1923). Inaccuracies in the Mathematical Literature. *Scientific Monthly,* 17, (3), 216-28.

[34] Monge, G. (1798). *Géometric descriptive. Leçons donnée aux école normales de l'an 3 de la république.* Paris: Baudouin, imprimeur du corps legislatif et de l'institut naional.

[35] Naudé, P. (1737). *TrigonoscopiæCujusdam Nova Conspectus,* in *Miscellanea Berolinensia ad incrementum scientiarum ex scriptis Societati Regiae Scientiarum.* Halae Magdeburgicae.

[36] Ozanam, J. (1684). *Récréations mathématiques et physiques.* Paris: Jombert.

[37] Petrilli S. J. (2009). *A Survey of the Contributions of François-Joseph Servois to the Development of the Rigorous Calculus.* Doctoral thesis: Columbia University Teachers College.

[38] Petrilli S. J. (2012). "Servois' 1813 Perpetual Calendar, with an English Translation." *Loci: Convergence,* **9**.

[39] Petrilli S. J. (2013). "Monsieur François-Joseph Servois," Canadian Society for History and Philosophy of Mathematics (pp. 60–71). *Proceedings of the 39th Annual Meeting*, Hartford, CT.

[40] Pickover, C. (2009). *The Math Book: From Pythagoras to the 57th Dimension, 250 Milestones in the History of Mathematics.* New York: Sterling Publishing Company, Inc.

[41] Poncelet, J. V. (1865). *Traité des propriétés projectives des figures*, 2nd ed., Paris: Gauthier-Villars.

[42] Rudé, G. (1955). The Outbreak of the French Revolution. *Oxford University Press on Behalf of the Past and Present Society* 8, 28–42.

[43] Servois, F. J. (1804). *Solutions peu connues de différens problèmes de géométrie-pratique; pour servir de supplément aux Traités connues de cette Science.* Metz: Chez Devilly/Paris: Chez Bachelier.

[44] Servois, F. J. (1811). "De principio velocitatum virtualium commentatis, in responsum quastioni ab illustrissima Academia Taurinensi, pro anno 1810, propositae, conscripta," in *Mémoires de l'Académie impériale des sciences de Turin* **18** pt. 2 (1809–1810), 177–244.

[45] Servois, F. J. (1813). "Calendrier perpétuel," *Annales de mathématiques pures et appliquées* **4**, 84–90.

[46] Servois, F. J. (1814). "Essai sur un nouveau mode d'exposition des principes du calcul différentiel," *Annales de mathématiques pures et appliquées* **5** (1814–1815), 93–140.

[47] Servois, F. J. (1814). "Réflexions sur les divers systèmes d'exposition des principes du calcul différentiel, et, en particulier, sur la doctrine des infiniment petits," *Annales de mathématiques pures et appliquées* **5** (1814–1815), 141–70.

[48] Servois, F. J. (1814). "Sur la théorie des imaginaries, Lettre de M. Servois," *Annales de mathématiques pures et appliquées* **4** (1814–1815), 228–35.

[49] Servois, F. J. (1817). "Mémoire sur les quadratures," *Annales de mathématiques pures et appliqués* **8**, 73–115.

[50] Servois, F. J. (1826). "Lettre sur la théorie des paralleles," *Annales de mathématiques pures et appliquées* **16**, 233–38.

[51] Sorrel, J. (1981). *Histoire de Saint-Marcellin (2nd ed).* Mines: Bené.

[52] Struik, D. J. (1948). *A Concise History of Mathematics.* New York: Dover Publications.

[53] Taton, R. (1972). "Servois," in *Dictionary of Scientific Biography*, 1972, C. C. Gillespie, Ed., New York: Scribner, XII.325-326.

[54] Tentomas, P. C. (2010). *Creation and Assessment of a Ten-Day Spatial Transformation Intervention in a Secondary School Geometry Class.* Doctoral thesis: Columbia University Teachers College.

[55] Vauban, S. (1691). *The New Method of Fortification: As Practised by Monsieur de Vauban, Engineer General of France, with an Explication of All Terms Appertaining to that Art.* English translation by Abel Swall. London: Printed for Abel Swall.

[56] Virgil. (1794). *The Works: Of Virgil, Translated Into English Prose, as Near the Original as the Different Idioms of the Latin and English Languages Will Allow.* Translated by Joseph Davidson. London: Printed by assignment, from J. Davidson, for T. Longman.

[57] Young, John Radford (1830). *The Elements of Analytical Geometry: Comprehending the Doctrine of the Conic Sections, and the General Theory of Curves and Surfaces of the Second Order.* London: John Souter School Library, and J. and J. J. Deighton, Cambridge.

Index

Made in the USA
Monee, IL
04 November 2026

67384781R00075